Operator's Guide to Rotating Equipment

Operator's Guide to Rotating Equipment

An introduction to rotating equipment construction, operating principles, troubleshooting, and best practices

Julien LeBleu, Jr. and Robert Perez

AuthorHouse™ LLC
1663 Liberty Drive
Bloomington, IN 47403
www.authorhouse.com
Phone: 1-800-839-8640

© 2014 Julien LeBleu, Jr. and Robert Perez. All rights reserved.

No part of this book may be reproduced, stored in a retrieval system, or transmitted by any means without the written permission of the author.

Published by AuthorHouse   05/19/2014

ISBN: 978-1-4969-0868-1 (sc)
ISBN: 978-1-4969-0768-4 (hc)
ISBN: 978-1-4969-0867-4 (e)

Library of Congress Control Number: 2014908373

Any people depicted in stock imagery provided by Thinkstock are models, and such images are being used for illustrative purposes only. Certain stock imagery © Thinkstock.

This book is printed on acid-free paper.

Because of the dynamic nature of the Internet, any web addresses or links contained in this book may have changed since publication and may no longer be valid. The views expressed in this work are solely those of the author and do not necessarily reflect the views of the publisher, and the publisher hereby disclaims any responsibility for them.

Table of Contents

Introduction xi
Chapter 1—Machine Types and Critical Components 1
Chapter 2—The Importance of Lubrication 37
Chapter 3—Inspection techniques available to operators
 and field personnel 54
Chapter 4—How to Inspect Process Machinery 70
Chapter 5—An Introduction to Compressor Operations 90
Chapter 6—Lubrication Advice for Operators 115
Chapter 7—More Machinery Best Practices 125
Closing Thoughts 142
Appendix A
    Useful engineering facts 143
    Useful Conversions 143
Index 147

Dedication:

We would like to dedicate this book to our wives for their constant support and encouragement.

Acknowledgements:

The authors would like to thank:

1. Elaine Perez for editing the book and providing valuable suggestions during the writing phase.
2. Drew Troyer and Bob Matthews for their valued reliability insights found in Chapters 6 and 7 of this book.
3. Charles Le Bleu for editing and support.
4. Carol Conkey for performing the final copy-editing.

Introduction

Rotating machines power our production facilities by safely transporting a wide variety of liquids, gases, and solids. Those of you who have worked in a production site for any length of time know that not all process machines are created equal. Some are more critical than others; some are small and some are very large; some spin fast and some turn very slowly. The great diversity in their construction and application can be daunting to a new operator and a challenge to the veteran.

When process machines fail catastrophically, bad things happen. These bad things are called consequences; but not all consequences are equal. For example, if a critical machine fails, the consequence can be the stoppage of an entire process or even an entire process facility, which may have a significant cost related to the outage if it is lengthy. Another consequence associated with machinery failure is the release of a dangerous process fluid if its seal fails. A final example is the cost of a major machine failure. If an internal mechanical failure is not detected early, more and more secondary damage will be incurred, eventually resulting in a hefty repair bill. Catastrophic failure can cause serious safety and environmental issues as well.

Figure 1A depicts how machines tend to fail. On the left, a machine is operating at an optimal, or 100% condition; pressure, flows, and temperatures are normal throughout the machine. Then maybe a bearing begins to fail or an impeller fouls causing vibration. For whatever reason, the machine's condition begins to degrade. As the problem continues, we may begin to experience vibration and later to hear noise. If the machine is allowed to degrade further, we will start to sense heat and then see smoke. Finally, if we continue to keep running this ill-fated machine, it will fail catastrophically and stop suddenly. Dire consequences, such as a product release or fire, could also occur.

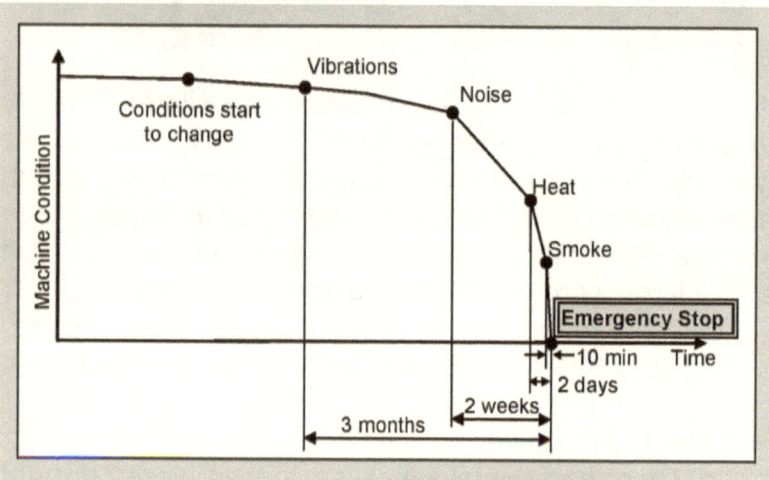

Figure 1A—How Machines Fail

This is no way to run a production facility! The goal of every operator should be to detect machine failures or degrading machine support systems as early as possible. Early detection of unwanted machine conditions, which is the central theme of this book, is the safest and most cost effective philosophy to employ. Identifying a problem before major damage occurs allows more time to plan the required repair or adjustments in order to minimize downtime and repair costs. An operator trained to identify degrading machine conditions before serious damage can occur is worth his or her weight in gold.

We all know that operating personnel are constantly being bombarded with more regulations, increasing training requirements, incessant paperwork, and that the number of operators per machine is dropping every day. This means the number of "eyes" available to watch critical equipment is constantly diminishing. But this is not a hopeless situation. This challenge to do more with less is driving production organizations to find more and more efficient ways of conducting business so that profits do not come at the expense of safety and the environment.

One such way to improve operational efficiency with little or no additional cost is to infuse an equipment reliability attitude into the organization from the ground up. Operators are the backbone of this reliability effort. Without committed, well trained operators reliability programs, no matter how well designed, are doomed to failure. To reap the full benefits of an equipment reliability program each and every operator must endeavor to:

- Gather data for troubleshooting and potential root cause failure analysis
- Identify symptoms of potentially damaging machine distress that can be rectified through adjustments, such as alignment, lubrication, balance, and changes in operating conditions
- Identify machine failures in their early stages so that secondary damage is avoided and ample time is available to plan proper repairs
- Maximize rotating equipment reliability by continuously striving to lessen the time between machine repairs

- Strive to operate machinery reliably and efficiently with the benefit of cost control for the owner.
- Protect life and limb
- Protect the environment

All site managers will agree that profit is meaningless if it comes at the cost of an injury or an environmental incident. It is prudent for an operator representative to be included in all capital work for input of a practical nature i.e. valve heights, orientation, and operability of equipment. Operators who give truthful and significant information on equipment and processes must not be disciplined for doing so. To do so would eliminate significant data and information in the future.

Working around process machinery equipment for a long time one can become accustomed to having it work as it should. Some machinery can even become "invisible" if it is out of sight or very reliable. It is easy to become oblivious to potential problems, especially if they develop gradually. The simple techniques of touching, listening, and visually inspecting machinery while on rounds or even just passing by will ensure the best possible life for machines and drive costs down by catching problems early before they develop into costly failures.

The key to detecting problems early is to know what "normal" is. The operator is in the ideal position to know how a piece of equipment is supposed to sound, what pressures are normally produced, and what the equipment feels like when it is doing its job.

The goal of this book is to present proven techniques that will enable rookie and veteran operators alike to detect problems early, eliminate major outages and control maintenance costs. To achieve this goal we will explain the basics of lubrication systems, bearings, drivers, seals and sealing systems for centrifugal and positive displacement pumps as well as turbines, centrifugal compressors and reciprocating compressors. We will then present common sense inspection methods for centrifugal and positive displacement pumps, gear boxes, motors, heat exchangers, and turbines.

After you feel you have mastered the first four chapters, you can move on to more advanced material found in the last three chapters which include: "An Introduction to Compressor Operations", "Lubrication Advice for Operators", and "More Machinery Best Practices".

Julien LeBleu and Robert Perez

Chapter 1
Machine Types and Critical Components

Drivers, Speed Modifiers, and Driven Machines:
Process machinery is typically composed of a group of sub-elements that convert one type of energy into another until it is finally transferred into a useable form of fluid power within a process. Here is a simple flow chart showing how power flows through a machine train.

Energy (in) ➔Driver➔Speed Modifier➔Driven Machine➔Process Fluid Power (out)

Machine train sub-elements are normally interconnected using flexible components called couplings. Figure 1.1 illustrates a simple machine train comprised of an electric motor directly coupled to a centrifugal pump.

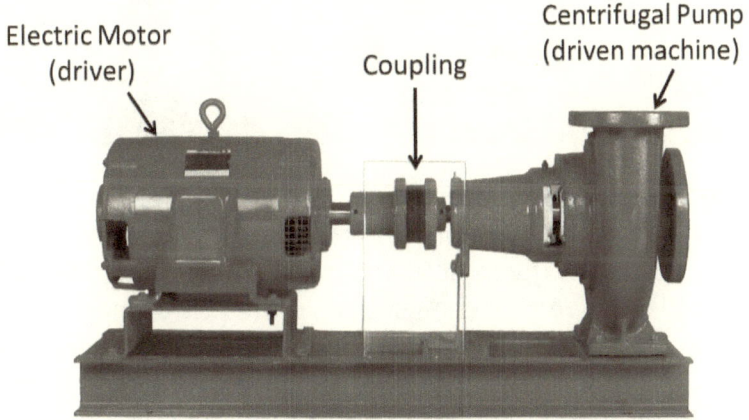

Figure 1.1—An electric motor coupled to a centrifugal pump

Energy, such as electrical power, steam power, or fuel gas, is first converted into rotational output power. The speed of the driver output shaft may be increased or decreased by a speed modifier, i.e. gearbox or pulleys, depending on the requirement of process machine being driven. Finally, the output speed from the speed modifier powers the driven machine that produces fluid power in the process. Table 1.1 contains common designs for driven machines, drivers, speed modifiers, and combination machines.

Table 1.1—Common types of process machinery elements

Driven Machines	Drivers	Speed Modifiers	Combination Machines
Centrifugal compressors	Electric motors	Gearboxes	Turbo expanders
Reciprocating compressors	Steam turbines	Hydraulic speed modifiers	Turbochargers
Rotary compressors	Gas turbines	Sheaves and belts	
Centrifugal pumps	Reciprocating engines		
Reciprocating pumps			
Rotary pumps			
Fans			

Driven Process Machines

The purpose of a driven process machine is to deliver a given process fluid, at a given flow and pressure, to specific points in a process. Driven machines receive the power input from a driver or speed modifier and convert it into fluid power at the process machine's discharge flange. All driven process machines are composed of an input shaft, a casing to contain the process fluid, a suction nozzle for input flow, a discharge nozzle for output flow, bearings to support the rotor (or rotors), and one or two end seals to prevent process leakage into the atmosphere.

There are many different designs employed to convert rotary power into fluid power.

Process machines that move and compress gases are called compressors or fans and process machines that move liquids are called pumps. There are too many designs employed in the process industry for us to cover them all adequately here. Instead of covering all design types, this chapter will concentrate on centrifugal and positive displacement pumps. Compressors will be covered in Chapter 5.

Centrifugal pumps
Centrifugal pumps are one of the most common types used in industry. Figure 1.2a shows a generic, single stage centrifugal pump and Figure 1.2b illustrates a multistage centrifugal pump. These pumps can utilize either open or closed impellers and may have single or multiple stage designs. Centrifugal pumps utilize Bernoulli's principle to develop pressure (see *Bernoulli's Principle Explained* below) by first increasing the fluid velocity inside an impeller and then decreasing the fluid velocity in the discharge nozzle. These pumps consist of a shaft with bearings for support and an impeller as well as a pump casing. To prevent leakage from the pump casing to the atmosphere, most pumps employ packing, single or dual mechanical shaft seals.

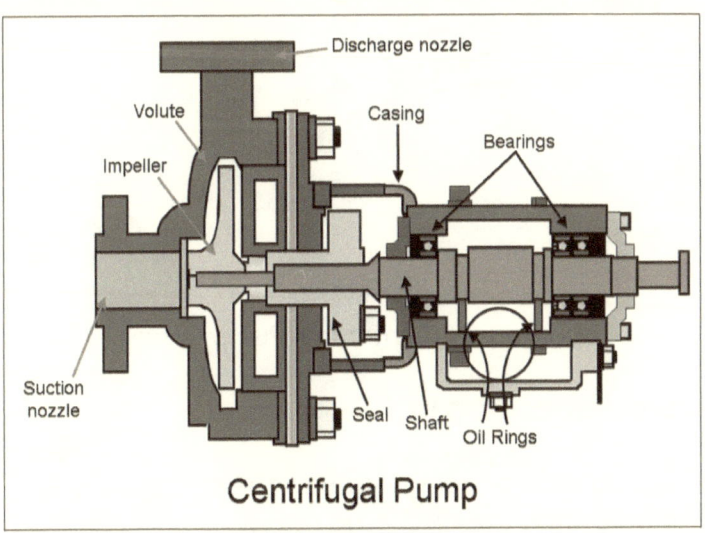

Figure 1.2a—Single Stage Centrifugal Pump

Figure 1.2b—Multistage Centrifugal Pump

Centrifugal pump performance is typically presented graphically with a series of curves similar to the group shown in Figure 1.3. The manufacturer usually provides curves that describe how flow, differential head, net positive head required, and efficiency change with pump flow.

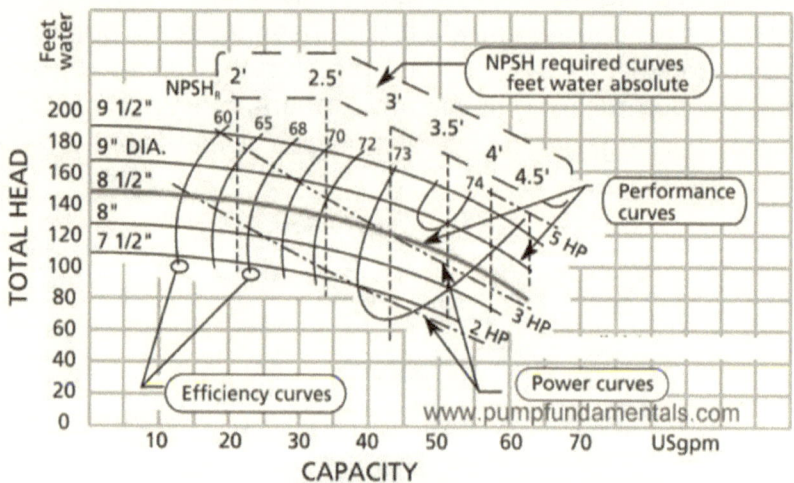

Figure 1.3—Centrifugal Pump Curves

Useful centrifugal pump facts:

- The suction or inlet nozzle to the pump is always bigger than the discharge nozzle.

- If a centrifugal pump has more than one impeller inside of it is called a multistage pump. If it has, for example five impellers in it, then it is a five stage pump.

Series and Parallel Operation

Centrifugal pumps may be operated in a series or parallel (see Figure 1.4) configuration. When operating pumps in series, the pressure is increased across each pump, but the flow through each pump is identical (minus any minor flow losses due to leakage). When operating in parallel, the pressure rise on each pump is identical, but the total flow is increased. However, the overall flow is not doubled with two pumps operating in parallel because of "system head" or pressure. The easiest way to understand system head is to remember that the discharge pipe size stays the same diameter and therefore tends to restrict the higher flow generated by two pumps operating in parallel. This bottleneck effect means that two pumps operating in parallel will always deliver less than twice the flow that one pump can deliver.

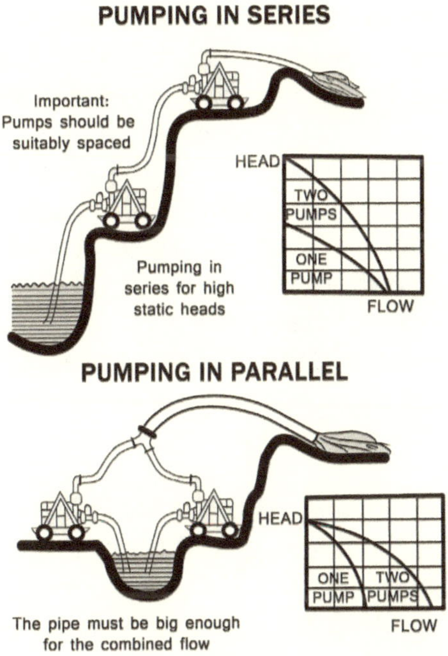
PUMPING IN SERIES
Important:
Pumps should be
suitably spaced
HEAD
TWO
PUMPS
ONE
PUMP
FLOW
Pumping in
series for high
static heads
PUMPING IN PARALLEL
HEAD
ONE
PUMP
TWO
PUMPS
FLOW
The pipe must be big enough
for the combined flow

Figure 1.4—Pumps in series and parallel

Bernoulli's Principle Explained

There are three physical forms of a fluid energy: Elevation energy, pressure energy, and velocity energy. The higher a liquid is stored, like water in a water tower, the greater its potential energy. The greater a fluid stream's pressure, the greater it's potential to do work. Similarly, the greater the velocity of a stream of fluid, the higher its capability to do work. As a fluid flows down a pipe, ditch, or river, there is a constant interaction between these three forms of energy.

The interplay of the three forms of fluid energy in a flowing stream is governed by Bernoulli's principle. Originally formulated in 1738 by the Swiss mathematician and physicist Daniel Bernoulli, it states that the total energy in a steadily flowing fluid system is a constant along the flow path. An increase in the fluid's speed must therefore be matched by a decrease in its pressure, i.e., energy is always conserved in a fluid stream.

- This principle explains why a moving stream of liquid or gas exerts less pressure than if it were at rest. Bernoulli's Equation can be used to approximate flow parameters in water, air, or any fluid stream that has very low viscosity as long as the fluid is assumed to have these qualities: fluid flows smoothly fluid flows without any swirls (which are called "eddies") fluid flows everywhere through the pipe (which means there is no "flow separation") fluid has the same density everywhere (it is "incompressible" like water)

In basic terms, the Bernoulli principle states that:

- At a constant velocity, if the elevation of a fluid stream increases, the pressure in the stream will decrease.
- At a constant velocity, if the elevation of a fluid stream decreases, the pressure in the stream will increase.

- At a constant elevation, if the velocity of a fluid stream increases, the pressure in the stream will decrease.
- At a constant elevation, if the velocity of a fluid stream decreases, the pressure in the stream will increase.

The last rule is the reason centrifugal pumps and centrifugal compressors are able to convert a high velocity flow from a rotating impeller into high pressure.

We can use the example of how an airplane wing works to better understand Bernoulli's principle. Because of the wing's shape, air moving over the top of the wing must travel farther and therefore faster than the air traveling across the bottom of the wing. This difference in air velocities results in a lower pressure across the top of the wing than under the wing, resulting in a net upward force that can lift an airplane. An easy way to remember Bernoulli's principle is as follows: If the fluid's velocity increases the pressure decreases; and if the fluid's velocity decreases the pressure increases. These two relationships are inversely proportional. (Figure 1.5)

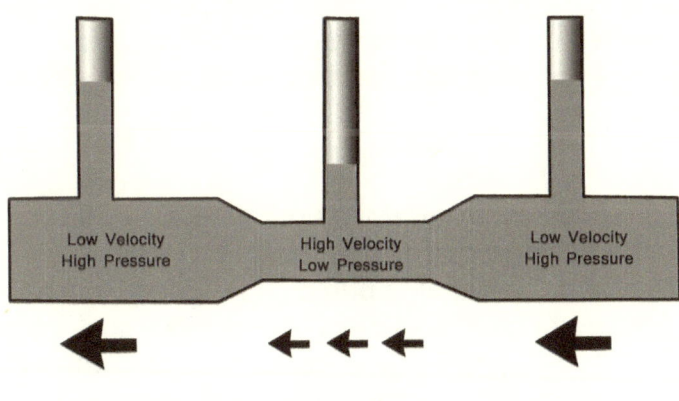

FLOW

Figure 1.5—Venturi (constricted pipe), demonstrating Bernoulli's principle. As the fluid travels from right to left it starts at a low velocity and high pressure in a wide portion of the venturi. As the velocity increases in the narrow portion, the pressure drops. As the velocity decreases in a wider portion the pressure again increases.

Julien LeBleu, Jr. and Robert Perez

Troubleshooting Centrifugal Pumps

Here are some useful relationships to remember when dealing with centrifugal pumps:

- If the pressure on the discharge pressure gauge is increasing, the flow is likely decreasing.
- If the flow is increasing, the horsepower required by the pump is increasing, which should be reflected by increasing amps or kilowatts.
- If the fluid viscosity is increasing or getting thicker, the discharge pressure will fall and the horsepower required by the pump will increase.
- If the flow is increasing, the net positive suction head required (NPSHR) will need to increase to prevent cavitation (see definition below).

> **Cavitation** is a serious operating condition that sounds like gravel is passing through a pump. This unique "gravelly" sound associated with cavitation is due to the fact that vapor cavities or bubbles are continually forming and then collapsing in the pump's inlet. If cavitation is not addressed and corrected quickly, the internal components of a centrifugal or positive displacement pump may be seriously damaged.

Cavitation is caused by either 1) operating a pump too close to the boiling point of the liquid at its suction or 2) by trying to pump more than a pump is designed to handle. Pump designers use a term called net positive suction head (NPSH) to determine whether there is enough suction pressure to prevent cavitation. Typical net positive suction head requirements can be seen in Figure 1.3 (Centrifugal Pump Curves), where they are depicted as dotted vertical lines, labeled: 2', 2.5', 3', 3.5', 4', 4.5'. 2' means that at this particular flow, two feet of liquid suction head is required to prevent cavitation, 2.5' means that at this particular flow, two and a half feet of liquid suction head is required to prevent cavitation, and so forth.

As an operator, you simply need to remember that the higher the flow rate out of a pump the greater the suction pressure must be to insure that cavitation does not take place. An operator is in the position to control some parameters that will affect NPSHA. For example, if the pump is taking suction from a tower or tank, either the liquid level or the pressure can be increased as a means of stopping or reducing pump cavitation. Reducing the pump flow can have a positive effect on the NPSHA as well and may stop cavitation.

Responding to Pump Problems
When you detect a pump problem and decide to act, always list the symptom or symptoms that were noted on the work request, such as not enough pressure or flow, a seal is leaking, the drive motor is tripping off line, there is high vibration on the pump, etc. Do not list what should be done about the suspected problem on the work request. Comments such as: fix the pump, change the pump, replace the pump, etc. are not helpful to the maintenance planner. Comments such as: Seized, vibrating, low pressure, tripping off line, etc. are helpful. Always, allow the maintenance crew to investigate the situation and perform what they think are the necessary corrections, adjustments or repairs.

Problems that overhauling equipment will not solve

- Plugged suction strainers.
- Plugged discharge strainers.
- Air leaking into a suction flange.
- Check valve on standby pump leaking.
- Pump turning backward.
- Wrong pump speed.
- Bypass valve open requiring too much flow.
- Blockage downstream of the pump.

- Too much flow being required from the pump.
- All of these should be checked before writing a work request for repair.

Positive displacement pumps

Positive displacement pumps (see Figures 1.6 and 1.7), unlike centrifugal pumps, theoretically can produce the same flow at a given speed (RPM) regardless of the discharge pressure. Thus, positive displacement pumps are constant flow machines. However, a slight increase in internal leakage as the pressure increases prevents a truly constant flow rate. As the pump is turned faster, more liquid is ejected from the pump. There are only two ways to increase output where a positive displacement pump is used—turn it faster or buy a bigger pump.

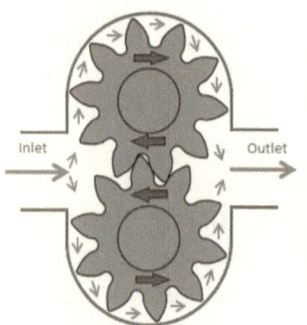
Figure 1.6—Gear Pump

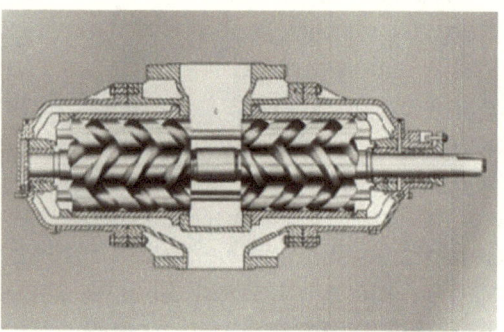
Figure 1.7—Screw Pump

In this type of pump, flow on the discharge side of the pump must never be stopped because discharge pressure will continue to build until something is damaged. There are many types of positive displacement pumps, but regardless of how a particular unit is configured, the results are the same. As the pump turns, the liquid inside it is trapped and must have a place to go. Positive displacement pumps have the same size inlet and outlet. On many of these pumps the suction and discharge can be changed by turning the pump in the opposite direction. Without a functional relief device on the discharge of the pump, catastrophic damage could occur.

Operator's Guide to Rotating Equipment

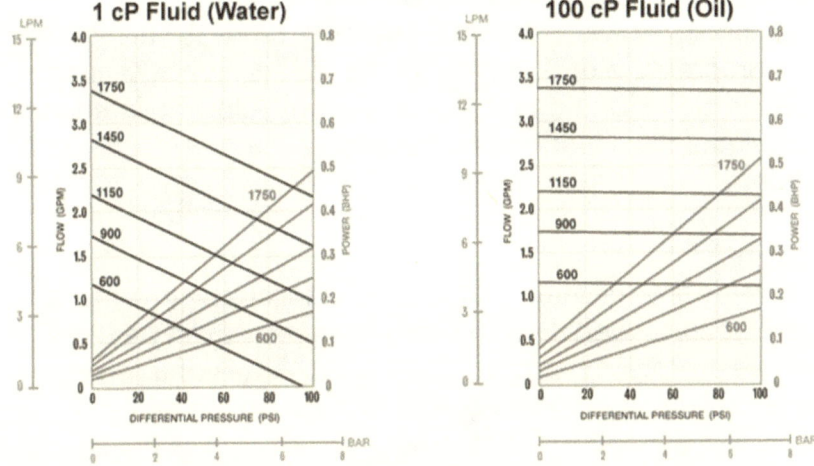

Figure 1.8—Positive Displacement Pump Curves

Positive displacement pumps also have performance curves (see Figure 1.8 above), but they tend to look quite different than centrifugal pump curves.

Useful relationships to remember when dealing with positive displacement pumps include:

- If the discharge pressure is rising, the required horsepower is also rising.
- If the pump is required to deliver more flow (e.g.— gallons per minute, barrels per day, etc.), then the speed of the pump must be increased or a larger pump is required.
- If the viscosity of the fluid is increasing, then the horsepower required to pump the fluid is also increasing.
- If the pump flow is increasing, then the net positive suction head required to prevent cavitation must also increase.

These basic relationships can help to troubleshoot a problematic positive displacement pump. When you detect a pump problem and decide to act, always list the symptom or symptoms that were noted on the work request, such as 'not enough pressure or flow', 'the outboard seal is not sealing', 'there is high vibration on the pump', etc. Allow maintenance to evaluate the situation and determine the proper course of action.

Drivers

A driver is any machine that takes electrical, steam, or fluid energy and converts it into rotary power that can be used to drive a process machine. The key attributes of a driver are reliability and efficiency. Various types of drivers are employed depending on the requirements of the driven machine and the available energy sources. For example, if electrical power is available and a constant speed is required for a driven machine, then an electrical motor can be used. If steam is available and variable speed is required to drive a process machine, then a steam turbine can be utilized. Steam turbines can be very efficient by driving a piece of equipment such as a pump and reducing steam pressure at the same time, eliminating the need for a steam let down or control valve.

If there is a difference between the design speed of the driver and its driven machine, a speed increaser or decreaser may be required to properly couple them. Gearboxes and a system of sheaves, belts, or chains can be used to transmit the required power and the required rotational speed. Size, cost, efficiency, and reliability are all factors an engineer considers when selecting the type of speed converter to be used.

Electric Motors

Electric motors are the most common type of rotating equipment driver in most process facilities. They are used not only to drive pumps but also to drive many other types of equipment found in industry. Figure 1.9a below shows a totally enclosed, fan-cooled (TEFC), electric motor, which consists of two major components (see Figure 19b)—A rotating element, called the rotor, and a stationary housing encircling the rotor, called the stator. A stator is composed of a set of stationary electrical windings that generate a rotating magnetic field created by the three phase electric power connected to them.

Figure 1.9a—Electric Motor with TEFC Enclosure

Figure 1.9b—Electric Motor Cutaway

Three phase power, commonly used throughout industry, is composed of three circuit conductors carrying three alternating currents (of the same frequency but 120 degrees out of phase). From an operations point of view, three phase power means that if any two leads connecting the motor are moved to a new mounting location externally at the connection to the power source, the direction of the motor will be reversed. Remember that, because many types of equipment *must* turn in a specific direction to perform properly, it is imperative that the direction of rotation of a newly wired electric motor be checked before installing the coupling between the driver and the driven machines.

As a general rule, if you touch a motor and it is too hot to keep your hand on it, it is likely that the motor is running heavily loaded. The fins on the outside of a motor are to aid with cooling and to stiffen the stator. These should be kept clean and clear of debris and insulation. Quite simply, the hotter a motor runs, the shorter its life will be.

There are many factors in determining what is too hot, such as the class of insulation and the load on that motor at a given time. The best analysis is done by knowing what normal temperatures are and by responding if a significant change in the operating temperatures is noticed. Additional information, such as the significance of the change, may mean getting help from an acknowledged professional.

Normally, a motor will not fail immediately due to higher operating temperatures. The effects of high temperature operation are cumulative and manifest themselves over time, resulting in a shortened service life. Large motors with an enclosure for placement outdoors need to be checked to ensure the screens or filters that protect the windings in the stator stay clean. If they become clogged, the temperatures may go up and the motor life may be significantly reduced.

Steam Turbines
Steam turbines, which can be likened to a child's pinwheel, are another common driver for industrial equipment (See Figure 1.10a). The turbine may have multiple "pinwheels" or stages with the wind replaced by steam. The shafts are sealed to keep the steam from leaking out. The seals preserve the heat and the treated water and keep the steam from going where it is not wanted. Badly leaking steam seals can result in

steam entering the bearing housings and where it condenses. The water will then return with the oil to the reservoir. Because the oil pumps draw near the bottom of the tank, eventually they will be circulating water, or a mixture of water and oil. We already know that a mixture of oil and water is a poor lubricant and therefore can lead to rapid bearing damage.

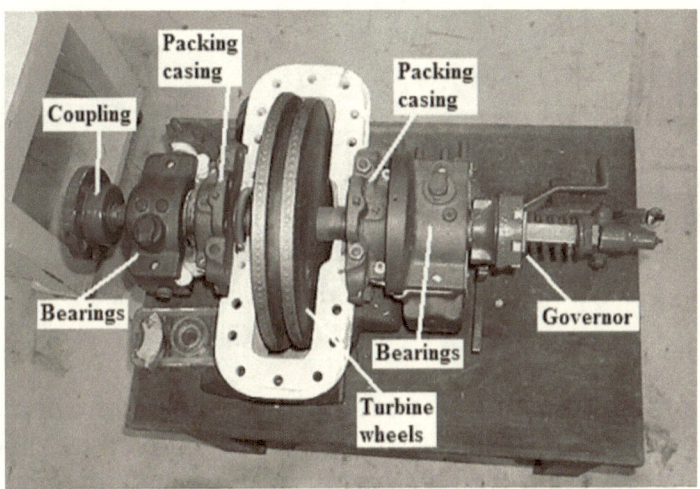

Figure 1.10a—Steam Turbine

If steam is seen leaking out of the turbine, the reservoir should be checked to ensure that large quantities of water are not accumulating there. The easy way to do that is to look for increasing oil level. Some turbines have cooling water being supplied to the bearing area. Touch the inlet and outlet water lines to determine if heat exchange is taking place or at least that flow is present. Watch for oil leaks, as the oil may collect in the insulation and come in contact with the hot piping or casing, causing a fire.

Another inspection point on these small turbines is the overspeed trip mechanism. The level should be well seated or it could fall at any time causing the turbine to trip and potentially shut the plant down. See figure 1.10b.

Figure 1.10b—Steam turbine with overspeed trip mechanism and valve

Notice if the steam turbine seems to vibrate more when it rains than when the day is sunny. The insulation may allow water to enter and cool the casing in an uneven manner, affecting alignment and how smoothly the turbine runs. Only operators are likely to notice these subtle sets of cause and effect associated with steam turbines.

Gear boxes

Gearboxes (see Figure 1.11) are often used whenever there is a large difference between the speed of the driver and the driven piece of equipment, high horsepower is required to be transmitted or input and output shafts need to be in different directions. A speed increaser is one where the input shaft is turning slower than the output shaft. A speed reducer is the opposite.

Operator's Guide to Rotating Equipment

Figure 1.11—Gearbox

From an operator's or casual observer's point of view, the most notable items when passing by a gearbox are the sound the gears are making and the feel of both temperature and vibration when touched. They need to be compared to the usual levels. Have they changed from what is accepted as normal and if so, "why". Always check foundation bolts by looking for oil being squeezed in and out between the base plate and the foot of the gear box. Look at the shims as well. Do they look like they have been squirming out? This is a potential sign of looseness or high vibration at some time. Occasionally check the foundation bolt or nut to ensure it is tight. This bolt could be broken off in the foundation (not easily detected), relaxing its grip on the rotating equipment. If, when trying to turn the bolt, it moves, then it is likely the stud in the foundation that is broken.

Useful Gearbox Facts:
The input and output shaft direction of rotation will depend on how many gear shafts are inside the gearbox. An even number of gear shafts equals opposite direction of rotation for input and output shafts; an odd number of gear shafts mean the same direction of rotation for input and output shafts.

Equipment Components
The overall reliability of process machines depends in large part on the reliability of their components, such as bearings

and seals. Each bearing and seal design comes with its own set of challenges. If the proper bearings and seals are selected and maintained, you can expect long service lives from your machines. The following is a brief explanation of the various types of bearings and seals commonly found in process machinery.

Bearing Types

Bearings play crucial roles in machine reliability and performance. Bearings maintain the position of rotors, so that internal clearances are maintained for efficient operation, while preventing the contact between the rotating and stationary parts. A bearing failure will lead directly to a machine failure.

There are many different types of bearing designs. Plain and rolling element bearings are two of the most common in process machinery.

Plain

Plain bearings, also referred to as journal bearings or bushings, represent a general category of bearings that are the simplest in terms of design. They consist of a bearing material that is usually stationary and a rotating shaft (see Figure 1.12). They may be lubricated by any of the methods of lubrication mentioned in the lubrication section (Chapter 2) of this book.

One type of plain bearing is called the pillow block bearing. Many of the larger ones have water connections going into them for cooling. Normally a water cooled pillow block bearing will have a water-in and out line on the same side of the bearing and a loop or "U" on the other side. Many times these lines are rubber hoses that can dry out and crack after years of exposure to the outside elements, allowing the cooling water to bypass the bearing and the oil to get too hot. Additionally, if poorly sealed where these connections enter the pillow block bearing, outside elements may enter the bearing oil. If a cooling water hose is leaking, it can contaminate the bearings lubricant by allowing cooling water to enter the bearing housing resulting in a bearing failure.

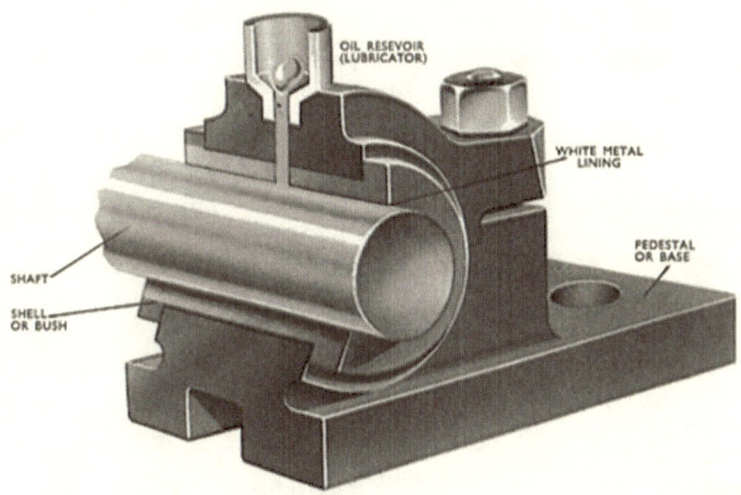

Figure 1.12—Cutaway of a plain (journal) bearing and shaft (Source: World of David Darling website)

Rolling element
Rolling element bearings fall into a large category of rotor support bearings that include ball, roller, and tapered roller designs (see Figures 1.13). They can be lubricated by either grease or oil, depending on the manufacturer's recommendations. If a rolling element bearing is known to be running very hot, do not put a water hose on the outside of the housing. The water could get into the lubricant, causing a premature failure. The cooling water on the outer housing will shrink the outer housing and may cause the bearing to seize immediately, or at least have unintended consequences.

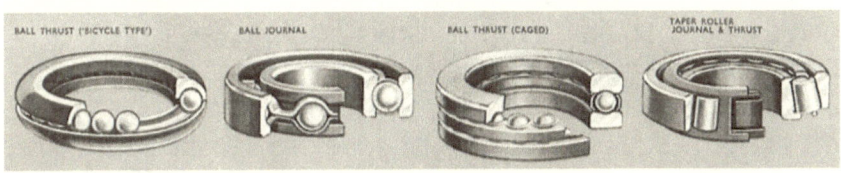

Figure 1.13—Different types of rolling element bearings.

Figure 1.14 contains a list of bearing types, categorizing them in terms of their construction. Plain bearings fall into two categories: metallic and non-metallic, with each of these types having many different design options. Similarly, rolling element bearings fall into two main categories: unmounted and mounted. Unmounted rolling element bearings can further be broken down as either ball or roller types. Mounted rolling element bearings can be broken down into design options, such as pillow block bearings, flanged units, cartridge units, and take-up units.

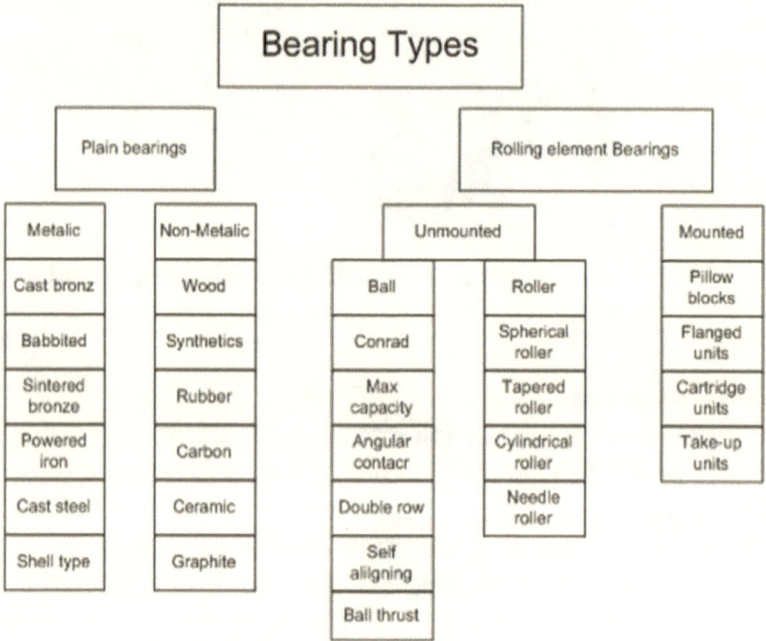

Figure 1.14—Types of bearings

The main takeaway from Figure 1.14 is that there are many types of bearing configurations available to machinery designers that utilize different materials and geometries. As an exercise, the reader can select a piece of equipment in a plant and determine the types of bearings it contains by talking to the people that maintain the equipment or by using the maintenance manual. It is beneficial for operators to understand what style of bearings a given machine uses in order to know how they must be maintained, and to be able to relate it to the information in the lubrication section in Chapter 2.

Pump Packing Rings and Mechanical Seals
Sealing a pump casing around the shaft periphery to prevent the pumped liquid from leaking into the atmosphere is a major concern. Through the years many seal designs have been invented and improved upon in order to ensure that product leakage rates do not exceed acceptable levels. Stricter and stricter leakage limits of regulated liquids have led to increasingly sophisticated seal designs. A wide range of designs satisfies most sealing requirements.

Effective pump sealing can be accomplished by using either packing or a mechanical seal. Years ago, most pump shafts were sealed using rings of soft packing compressed and held in place with a packing gland (see Figure 1.15). This type of shaft seal required a fair amount of leakage just to lubricate the packing and keep it cool. Later, the mechanical seal (see Figure 1.16) was developed to control leakage by utilizing two very flat sealing surfaces (one stationary and one rotating) inside the sealing cavity. The entire arrangement was designed to fit in the same volume of the original packing gland. Even though these mechanical seal faces also require some (very small) leakage across the faces, to cool and lubricate the seal faces, this leakage normally evaporates and is not noticeable. Most pump shafts today are sealed by means of mechanical seals. However, because of the delicate components used for this new sealing method and the requirements of the seal support system, mechanical seal failures are the greatest cause of pump down time.

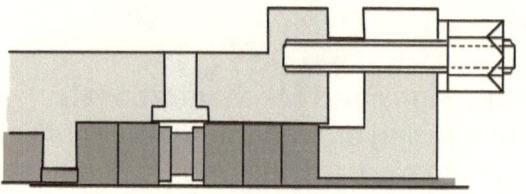

Figure 1.15—Packing Gland

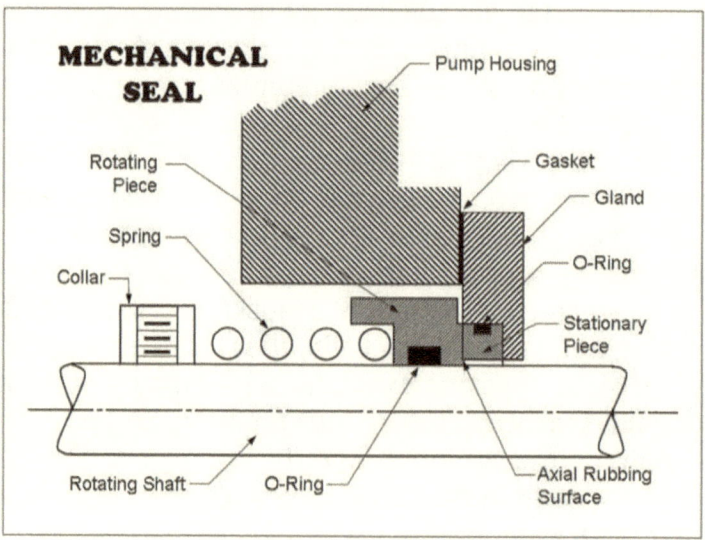

Figure 1.16—Mechanical Seal

With today's technology, mechanical seal designs can suit most application requirements, including temperatures up to 500 degrees F and shaft speeds of 3600 RPM or more, for pumps, through the selection of the proper combination of materials and auxiliary seal support systems. Seals can be ordered in balanced configurations to seal pressures above 200 psi, and designed with multiple sets of sealing faces to handle extremely high pressure applications.

The part of the pump casing that contains the mechanical seal or packing rings is called the stuffing box (or with a mechanical seal, the seal chamber). Horizontal end-suction pumps and vertical pumps have only one stuffing box, whereas horizontal between-bearing pumps, such as those used on cooling towers for high flow and low discharge pressure, are provided with two stuffing boxes one on each side of the pump casing.

Packing versus Mechanical Seals
Gland packing and mechanical seals have both been successfully used for sealing pump shafts. Countless engineering applications rely on these devices. Although mechanical seals and packing have similar functions, which one is used depends on budget, personal preferences, and more importantly, application requirements. A centrifugal pump moving either a flammable or hazardous fluid demands that a mechanical seal or seals be used to eliminate leakage to the atmosphere. On the other hand, there are applications where a packing is the more appropriate choice, such as cooling water service or other safe liquids such that that leakage will not result in safety or environmental violations. Some pros and cons of the different sealing methods are listed below.

Advantages of packing rings
1. Lower initial cost.
2. Very simple selection, installation, maintenance, and trouble-shooting.
3. The packing rings can act as an additional bearing to support the pump shaft reducing shaft deflection during extreme operations of the pump.

Disadvantages of packing rings
1. Require small amount of leakage for lubrication and cooling—therefore packing rings are unsuitable for use in liquids that are toxic, flammable, or otherwise hazardous or polluting.
2. Usually requires a stack of several packing rings for effective sealing. The act of sealing destroys

the shaft or sleeve of the pump, making more expensive repairs.
3. Easily wears out, resulting in more frequent replacement of packing rings and increasing the leakage from the pump.
4. Requires substantially more power to be applied to the pump in order to overcome packing friction on the shaft or sleeve.
5. Because leakage can increase over time, if not adjusted often the leakage can enter the bearing housing and contaminate the oil with failure of the bearing resulting.

Advantages of using mechanical seals
1. Reduced energy usage on the pump due to lower frictional drag than traditional packing.
2. Will not wear out a shaft or sleeve in the same manner as packing rings.
3. Near zero leakage possible with a mechanical seal; packing requires some leakage (usually visible) for proper lubrication.
4. Require less periodic maintenance than packing if properly applied and its support system maintained.
5. Specially designed mechanical seals can be applied to higher pressures and speeds than traditional packing.

Disadvantages of using mechanical seals
1. Less tolerant to shaft deflection and misalignment.
2. Less tolerant to dirty or contaminated liquid.
3. May require expensive seal support systems which need to be maintained.
4. More expensive than packing rings.

Packing Inspections and Adjustments
It is important to inspect pumps with packing frequently to ensure that there is leakage and that it is not excessive. For

most braided pump packing, liquid leakage is absolutely necessary to provide lubrication and cooling of the sealing surface and ensure the packing's long life. When adjusting pump packing, the goal is to arrive at the lowest acceptable leak rate while maintaining the life of the seal. Optimizing packing leakage rate requires the operator to have experience with the combination of packing material and liquid being pumped. Over-tightening can cause the packing to fail very quickly and is the most common cause of packing failure.

When making packing adjustments the guiding principle is to make adjustments that are proportional to the leakage rate. If the pump is spraying large amounts of leakage, then larger adjustments can be made until the leakage is controlled in the form of one to several drops per minute. After leakage is reduced, further adjustments should be small. One-sixth of a turn on the gland nuts (one flat) can have a significant effect. Several minutes should pass before the next adjustment is made. Adjust all gland packing nuts equally when adjustments are made. Follow the packing manufacturer's recommendation on adjustments. If the packing gland follower, the outside part with the adjusting nuts, moves very close to the packing gland notify maintenance as the packing is very close to requiring a change out. Some packings smoke on initial startup. Always check the manufacture's recommendations and information, so panic does not result.

Packing Leakage Paths
There are two principal paths that leakage can take out of the stuffing box (see Figure 1.17):

- *Inside diameter (ID)* leakage occurs along the interface between the packing and the rotating shaft or sleeve. This leakage cools and lubricates the dynamic surface and is necessary for most braided pump packing applications. Leakage can also be between the shaft sleeve and the shaft. It may look like the packing to

shaft or packing to sleeve leakage but it cannot be stopped by adjusting the packing.
- *Outside diameter (OD)* leakage exists when the liquid runs along the interface between the packing and the stuffing box bore. While it will have some cooling effect on the packing, this leakage is unnecessary because it will not serve as a lubricant for the spinning shaft and the stationary packing.

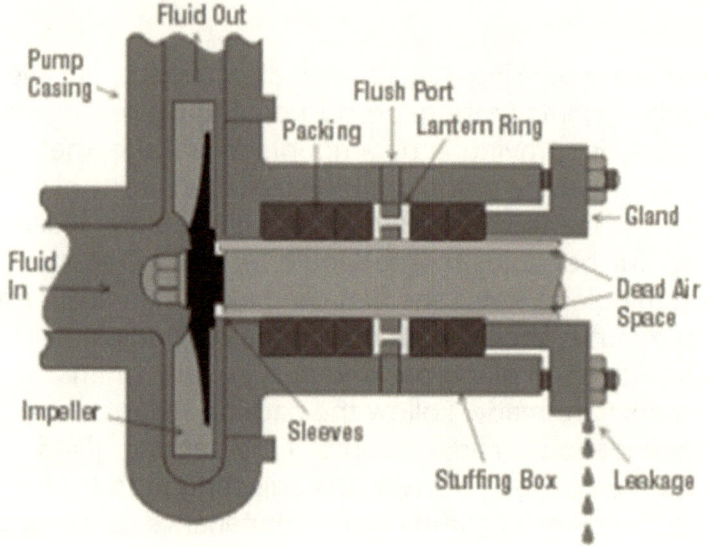

Figure 1.17—Pump packing leakage paths

Mechanical Seal Inspections

Most mechanical seals never reach the end of expected life and wear out. If a suitable seal design is properly installed into a cool, clean, and low-vibration environment, it could last for 10 or more years before wearing out. In reality, we rarely see a failure due to reaching the end of a seal's useful life. Unfortunately, we all have to live with imperfect seal designs and installations that often meet a premature end. For this reason, it is important that pumps with mechanical seals be inspected frequently so that small seal leaks are detected before the leak worsens. A seal is considered to have failed

Operator's Guide to Rotating Equipment

when leakage exceeds environmental or plant-site operating limits. This is the area, on mechanical seals, where the operator can have the maximum affect. It is important to run the equipment correctly—never dry or with cavitation. The seal support system must be inspected and kept operating. There are liquid levels and pressures to be maintained and cooling water to barrier fluids or seal life will be significantly reduced.

Understanding the mode of seal failure can lead to better troubleshooting and identification of the root cause of the seal failure. There are numerous possible reasons for a mechanical seal to fail prematurely. Several of these are listed below, with the factors under an operator's control tagged with an "**Operational Factor**" notation:

1. Incorrect seal design or material selection for the application temperature, pressure, speeds, and fluid properties.
2. Seal component abuse before installation—including chipping, scratching, nicking, or allowing to become dirty.
3. Erroneous installation—including assembly, seal setting, or placing of components in the chamber.
4. Improper start-up, dry-running, or failure of seal support systems. **Operational Factor:** Make sure you are using the proper start-up procedure to ensure the seal chamber is fully vented and filled.
5. Seal support system. **Operational Factor:** Make sure the proper seal flush rate is applied, and the proper level is achieved and maintained in the barrier fluid system. Ensure cooling water is being supplied to the cooling coils in the barrier fluid system and that heat is being exchanged between the inlet and outlet of the cooling water lines on the barrier system and the inlet and outlet from the mechanical seal. The upper line should be warmer than the lower line from the seal.
6. Improper pump operation. **Operational Factor:** Make sure the pump is operating at a safe flow rate and

is getting plenty of suction flow. Cavitation not only affects flow from the pump but can damage impellers and internal pump parts. It can be detected by a characteristic noise such as gravel or popping inside of the pump and the discharge pressure gauge needle is much more unsteady than normal.
7. Contamination of the sealing fluid with abrasive or corrosive materials. **Operational Factor:** Watch for upset conditions that can lead to a contamination of the seal flush.
8. Equipment has excessive shaft run-out, deflection, vibration, or worn bearings.
9. Worn out seal—the seal achieved its normal life expectancy. After a seal failure, always check to see how long the seal ran before it failed to see if the failure was premature or not.

Every mechanical seal leaks. Mechanical seals may be used as single or multiple seals, sometimes called double seals. With single seals it is the product that leaks to the atmosphere. With double seals the user gets to choose whether the product or a barrier fluid leaks to the atmosphere. Double seals are normally used in dirty or hazardous services, and they use a barrier fluid between the two mechanical seals to prevent hazardous materials from escaping. The fluid between the seals may be at a pressure higher or lower than the stuffing box pressure, depending on the application. Gas seals are a relatively new type of seal where a gas (usually nitrogen) is introduced between the two mechanical seals and seal faces as the barrier fluid. These seals are becoming very popular because, when properly applied, there is little to no contact between rotating and stationary parts and thus no wear. Nitrogen entering the pumped fluid normally does not offer a contamination problem.

A single mechanical seal may leak along one of five paths (see Figure 1.18).

1. Seal face leakage is visible where the shaft exits the gland but not under the sleeve or at the drain connections.
2. Dynamic secondary seal leakage is visible where the shaft exits the gland or at the drain connections.
3. Static secondary seal leakage is visible where the shaft exits the gland or at the drain connections.
4. Gland gasket leakage is visible at the gland-seal chamber interface.
5. Hook sleeve gasket leakage or cartridge sleeve secondary seal leakage is visible where the sleeve ends outside the seal chamber.

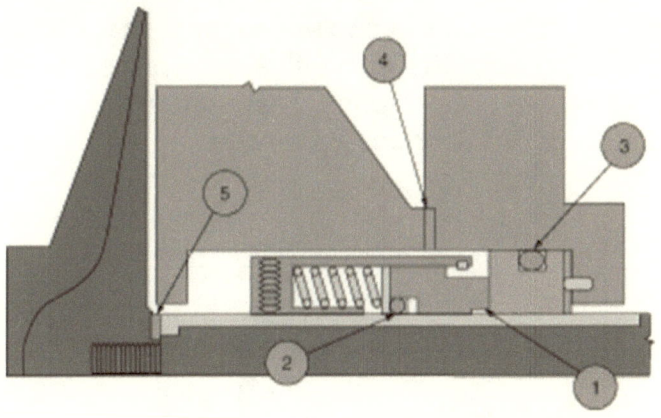

Figure 1.18—Possible seal leakage paths

The first four of these locations result in visible seal leakage. Leaks at locations 1, 2, and 3 will show up at the ID of the seal gland. A leak at location 4 will show up at the OD of the seal gland.

Mechanical Seal Flush Plans

To attain long lives from mechanical seals, think of them as bearings whose faces must 1) be lubricated, 2) kept cool, and 3) kept free of debris. In challenging sealing applications, external piping systems called seal flush plans may be required to ensure a healthy environment is maintained at the mechanical seal faces. Maintaining seal flush plans is one key area that operators can have the most effect. Below are two of the many possible seal flush plans. These plans are referred to by an American Petroleum Institute (API) number.

API 21 flush plan (Figure 1.19) takes flow from the discharge of the pump, cools it using a heat exchanger, and then returns it near the seal faces. There are two common problems encountered with this seal flush plan: 1) plugging in the exchanger, and 2) fouling on the water side of the cooler. During inspection rounds, always check to see that there is a temperature difference between the inlet and outlet lines. If the water outlet line is warm or hot, you can be sure cooling is taking place. These coolers should be back flushed at each shut down. If the seal flush cooler cannot be back flushed, let someone know so they can either be replaced or specially cleaned at the next opportunity. These are used in hot clean services such as hot condensate service.

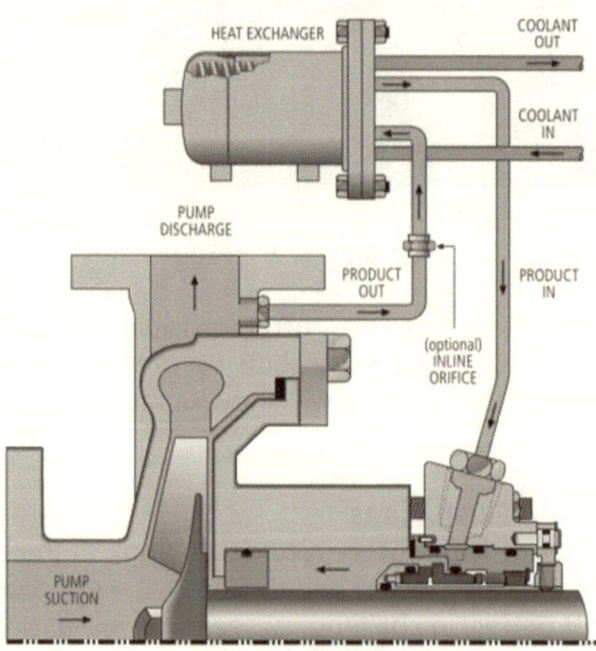

Figure 1.19—API 21 flush plan—Product flow from pump discharge goes through cooler before entering seal chamber

The API 32 seal flush plan (Figure 1.20) involves using a clean external flush to keep the mechanical seal faces cool and free of debris. A flush flow meter is recommended for this seal flush plan to provide a means of monitoring the flush rate. Place a mark or label on the flush rotor-meter so that you can tell what the correct flush flow should be. This arrangement may be used in hot or dirty service. The flush material being injected must be compatible with the pumped fluid and at a higher pressure than the stuffing box.

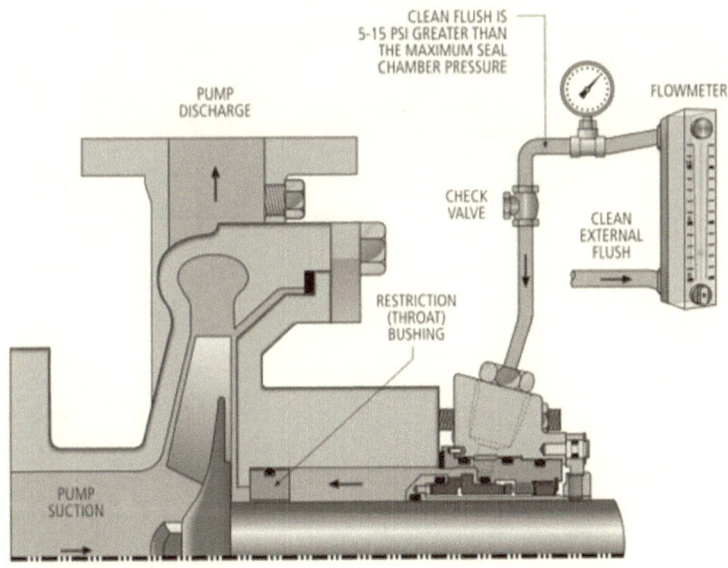

Figure 1.20—API flush plan 32. An in-line flow meter is recommended to provide a means of monitoring flushing flow

Heat exchangers

Process machinery tends to generate heat for several reasons: all bearings generate heat due to frictional losses, compressors generate heat through compression, and gearboxes create heat due to their inefficiencies. In order to maintain safe operating temperatures a way to remove heat created by process machines is often required. A heat exchanger is a device built for efficient heat transfer from one medium to another. It does not matter if the media are separated by a solid wall so that they never mix, or if they are in direct contact with the heat exchange surfaces. One of the most common types of heat exchangers in the chemical and petrochemical industry is the shell and tube (see Figure 1.21).

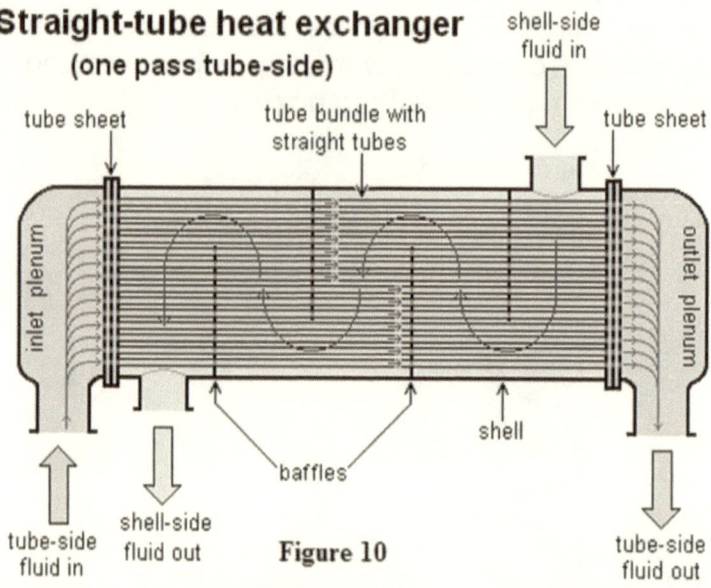

Figure 1.21—Shell and tube heat exchanger

Heat exchangers can transfer heat in three different ways. They can use conduction, convection or radiation. For example, a motorcycle cylinder is a heat exchanger that uses all three methods. Conduction is used from the cylinder and piston to the outside finned wall of the cylinder. When the motorcycle is moving and air is flowing over the fins convection is taking place. When stopped heat felt on the legs is from radiation. A common problem encountered in heat exchange equipment is fouling. Fouling can be caused when minerals in the cooling water precipitate out at high temperatures and plate out on metal surfaces. Eventually as scale builds up in a cooling water tube, it can reduce and even block off cooling water flow and dramatically reduce the exchanger's ability to remove heat. Figure 1.22 shows a severely fouled exchanger tube. In some cases biological life can attach to the inside of tubes and restrict or prevent flow in the tubes.

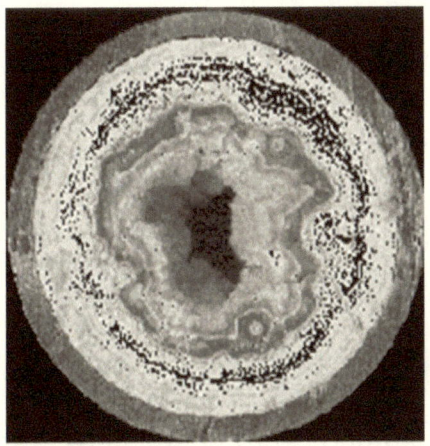

Figure 1.22—Severely fouled heat exchanger tube

Chapter 1
Review Questions

1. List two types of bearings.

2. List two types of drivers.

3. List two types of driven machines.

4. When operating centrifugal pumps in series, the pressure is increased across each pump, but the flow through each pump is _____.

5. _____ is a serious operating condition that sounds like gravel is passing through a pump.

6. On a centrifugal pump, what is likely happening to the flow if the pressure on discharge pressure gauge is increasing?

7. On a positive displacement pump, what is likely happening to the required horsepower if the viscosity of the fluid is increasing?

8. What are the two general categories of pump seals?

Chapter 1
Answers

1. List two types of bearings.
   - Plain
   - Rolling element

2. List two types of drivers.
   - Electric motors
   - Steam turbines

3. List two types of driven machines.
   - Centrifugal pumps
   - Positive displacement pumps
   - Centrifugal compressors
   - Positive displacement compressors

4. When operating centrifugal pumps in series, the pressure is increased across each pump, but the flow through each pump is **identical**.

5. **Cavitation** is a serious operating condition that sounds like gravel is passing through a pump.

6. On a centrifugal pump, what is likely happening to the flow if the pressure on discharge pressure gauge is increasing?
   - The flow is likely dropping.

7. On a positive displacement pump, what is likely happening to the required horsepower if the viscosity of the fluid is increasing?
   - The required horsepower is likely rising.

8. What are the two general categories of pump shaft seals?
   - Packing
   - Mechanical Seals

Chapter 2
The Importance of Lubrication

Lubrication is one of the most important aspects of any rotating equipment reliability program. Lubrication related failures are the most preventable type of machinery failures. With sufficient attention to details, they should be rare events. It is normally through the neglect of lubricants and lubrication systems that equipment fails prematurely. As a machinery operator your vigilance plays a central role in maximizing the useful life of a machine's lubrication system.

The lubricants and sealing methods available today have greatly improved the reliability of both pressurized, non-pressurized, sump and splash type lubrication systems. All machines require lubrication of their radial and axial bearings in order to maintain the rotor's position relative to the casing and to ensure reliable operation. Lubrication serves the following critical functions:

- Reducing wear by separating moving surfaces
- Reducing friction between the rotating and stationary components

- Absorbing shock
- Dampening noise
- Carrying heat generated by friction within the bearing
- Removing heat transmitted down the shaft from the process end of the machine
- Minimizing corrosion
- Keeping contaminates away from the bearing components
- Flushing contaminants away from bearings
- Acting as a sealing medium

Lubrication Regimes

There are four basic fluid film lubrication regimes encountered in machinery, listed here from thinnest to thickest:

1. **Boundary lubrication**, where surfaces are in contact with each other even though a lubricant is present (see Figure 2.1). This is generally an undesirable operating regime for a fluid film bearing since it leads to increasing friction, energy loss, wear, and material damage. If designed properly, most machines will experience boundary lubrication only during start-ups, shutdowns, and low speed operation. Special lubricants and additives have been developed to lessen the negative effects of this regime. When constant starts and stops create boundary lubrication the life of a machine is reduced.
2. **Mixed lubrication**, where surfaces are partially separated
3. **Elastohydrodynamic lubrication**, where two surfaces are separated by a very thin lubrication film. Elastohydrodynamic lubrication is employed in most rolling element bearings. Think of a balloon sitting on a table. As the balloon is pressed on the top the contact area on the table will increase, allowing the support of more load by reducing the unit loading. "Elasto" in the name means parts can deform

elastically; "-hydro-" refers to a liquid such as oil, and "-dynamic" means it is moving.
4. **Full fluid lubrication**, also called hydrodynamic lubrication, is where two surfaces are completely separated by a fluid film (see Figure 2.2). Full fluid lubrication is essential for the long-term reliability of high-speed, fluid film bearings. A full fluid film is formed when the bearing geometry, shaft speed, and oil viscosity combine to generate sufficient pressure to support the full bearing load. This is self-generating pressure that results in the complete separation of the shaft from the bearing.

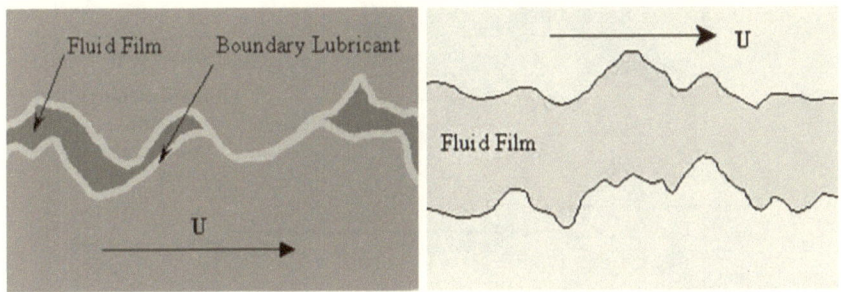

Figure 2.1—Boundary Lubrication

Figure 2.2—Full Fluid Lubrication

Types of Lubricants

Grease

Grease is a solid to semi-fluid mixture of a thickening agent (such as a chemical soap) and liquid lubrication. Some commonly used soaps are polyurea for electric motor applications and lithium for pillow block bearings and similar rolling element applications. Thicknesses range from 000 (semi-fluid grease) to 6 (block grease). It is typically used in lower speed applications (1800 rpm or less). The price of grease ranges from $0.20 to $100 per ounce. Always refer to the manufacturer's recommendation when selecting grease for your specific application.

Figure 2.3 shows a typical greased bearing with a "zerk"[1] fitting for grease addition. Remember: You should never mix greases without first checking with your lubrication department Mixing grease types can be considered the same as contaminating the lubrication—the result is either softer grease that allows lubricant to flow away from the application at a lower temperature or harder grease that decreases its ability to lubricate and flow properly.

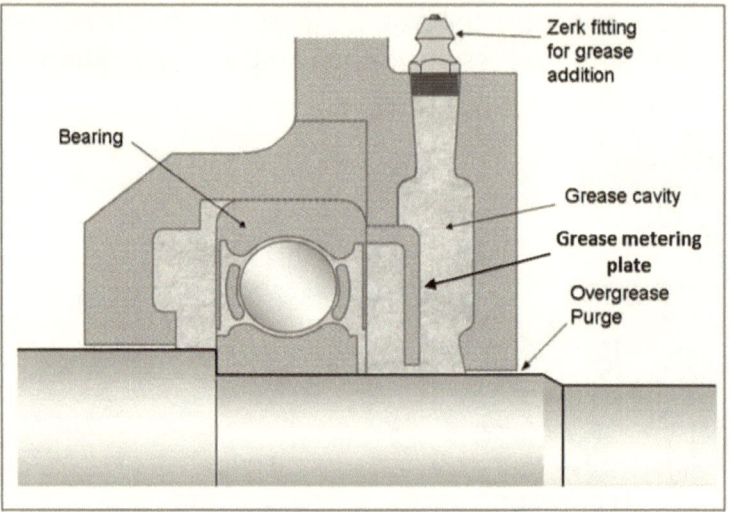

Figure 2.3—Double-shielded bearing with grease-metering plate facing grease reservoir

Greasing Tips

- Keep grease guns out of the weather (clean and dry).
- Clean the zerk fitting before adding grease; otherwise contaminants resting on the fitting could be pushed into the bearing with the fresh grease. Be sure no paint is on the zerk fitting, as it might prevent grease from getting to the bearing.

---

[1] Zerk fitting: a special check valve used to add grease to a bearing with a grease gun.

- Ensure the grease coupler is clean and free of hardened grease globs and dirt—before use. This can be done by capping the end or wiping excess grease off of the grease gun end and then operating the gun by pumping a small amount of grease out into a cloth to clean the grease passage way.
- Remove the vent plug to allow old grease to be expelled.
- Always follow re-greasing guidelines regarding the type, application interval, and amount of grease to apply during each application. Don't over grease!

Oil

Oil is a liquid lubricant frequently used in critical process machinery operating at speeds greater than 1500 rpm. Oil can be splashed onto rolling element bearings by either slingers/flinger rings (see Figure 2.4a) or oil rings. For splash lubrication to work properly, the correct oil level must be maintained in the bearing housing. The most common means of maintaining this level is through the use of a constant level oiler (see Figure below). Oil can also be used to lubricate journal bearings by means of slinger rings or pressure feed lubrication.

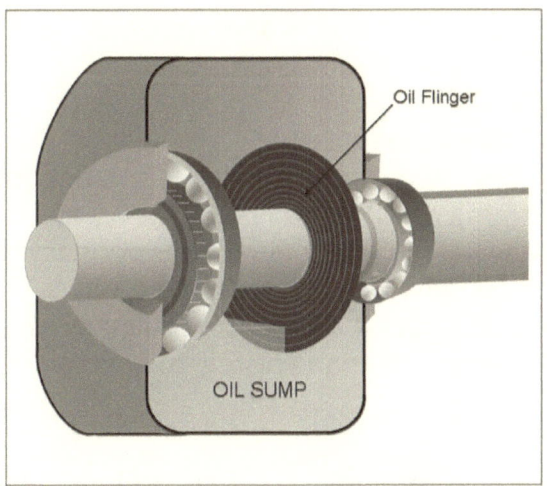

Figure 2.4a—Oil Flinger Ring

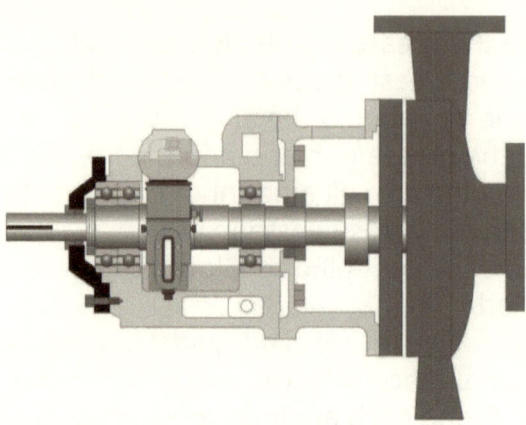

Figure 2.4b Centrifugal Pump with a Constant Level Oiler

Oils fall into three main categories:

1. *Animal/vegetable* oils which are not commonly used in industrial machines. Many of these are food grade and used in food applications.
2. *Mineral* oils are the most commonly used industrial lubricants due to their relatively low cost. They have moderate oxidation and viscosity stability, and cost $12 per gallon or more.
3. *Synthetic* (man-made) oils are used in the more demanding machine applications, cost $32 per gallon or more, and have very good oxidation and viscosity stability.

Types of lubrication systems
Machinery lubrication can be applied in a variety of ways depending on the equipment design, speed, and operating conditions. The machine's manufacturer normally determines the type and quantity of lubrication required. As stewards of equipment, it is your job to follow the manufacturer's lubrication recommendations closely to ensure reliable operation.

The most common ways lubrication is supplied to process machinery are:

Splash Lubrication—This type of lubrication system usually consists of a reservoir of oil and some part of the spinning shaft and attachment or the rolling elements of the bearing that touches the oil causing it to splash, allowing lubrication to take place. There is normally a place on the machine to check the oil level while it is in operation. Ensuring oil is always at the correct level is critical.

Ring Lubrication—This is a form of splash lubrication, accomplished by use of a large ring, usually brass, that rides on a turning shaft. The ring dips down into the oil reservoir and, by viscous drag, brings oil up onto the shaft where it is distributed along the shaft to the bearing. As stated above, there should be a place on the machine to check the oil level. During inspections, it is critical for operators to verify that bearing housing oil levels are at the correct level. There is also usually an inspection port (see Figure 2.5) that can be opened or a plug that can be removed in order to look at the oil ring while the piece of equipment is running. Inspection of the oil ring in service should be viewed often to ensure the ring is turning; if the ring is not turning, lubrication has ceased.

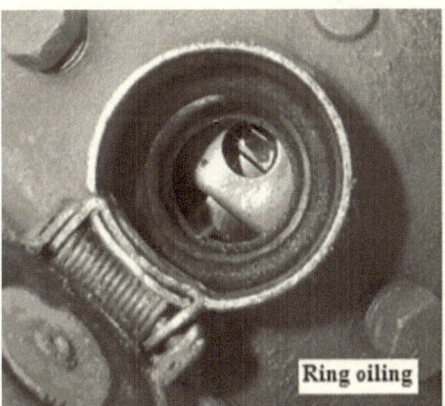

Figure 2.5—An Opened Ring Oil Inspection Port

Circulating Oil Systems—This type of lubricating system usually contains a reservoir, a pump, a filter, and may or may not have a heat exchanger (see Figure 2.6). This type of system supplies oil to the lubricated item at very low (essentially zero) pressure. It is used when a controlled flow of clean lubricant is necessary in one or more places which may or may not be at the same level as the reservoir. This is very similar to a forced-feed lubrication system, but uses its pump to simply circulate oil. Frequently check the level in the reservoir and the color of the lubricant. If there are changes, the question "Why?" should always be asked.

Figure 2.6—Circulating Oil System

Forced Oil Lubrication—This system is similar to a circulating system, but operates at an elevated lubrication supply pressure. It usually has a regulating valve to maintain pressure on the system, and utilizes coolers, multiple pumps, a pressure regulator, auto-start of the stand by pump, filters, and a reservoir (see Figure 2.7 below).

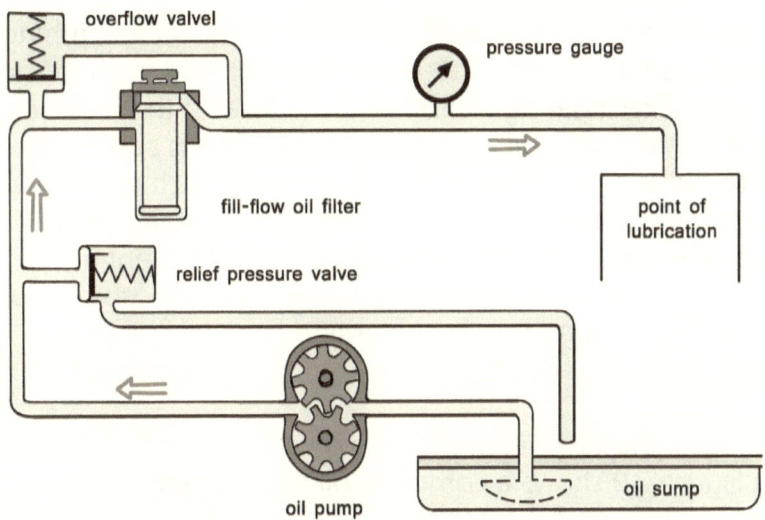

Figure 2.7—Basic Forced Lubrication System

This type of system is designed to remain at or above the design pressure continuously or the equipment it is lubricating will fail prematurely. There are usually safety switches that will cause the lubricated equipment to shut down if the level in the tank falls below a set amount, or if the pressure in the system becomes too low. In addition to checks of the circulating system, additional checks should be made, such as "Are both the main and auxiliary oil pumps running?" If they are both running, ask yourself some additional questions: "Why? What is the system pressure? Is it normal?" Touch the discharge side of the relief valve on the discharge of the pump. Is it relieving? It should not be.

Oil Mist Lubrication—A type of lubrication that is formed of oil particles 1.0 to 3.0 microns in diameter suspended in a current of air, consisting of 1 part oil to 200,000 parts air. This mixture is not a volatile organic compound (VOC); therefore, there is no risk of explosion or combustion. When the oil mist arrives at the bearing housing, it must pass through a reclassifier which, producing turbulence, causes the mist to coalesce into larger oil particles that are capable of "wetting" and lubricating the bearings.

Typical oil mist lubrication systems consist of a tank, tubing to each item being lubricated, an atomizer, and various safety devices relating to flow and the level in the reservoir. There are two general types of systems: One is *pure mist*, where there is no lubricant in a reservoir in each piece of equipment. The other type is a *purge mist* system, where there is a level in the lubricated piece of equipment and oil mist fills the space above the reservoir. Bearings lubricated with pure mist tend to run cooler and have longer service lives.

Figure 2.8 shows a bearing housing with pure mist lubrication and Figure 2.9 shows a complete oil mist system.

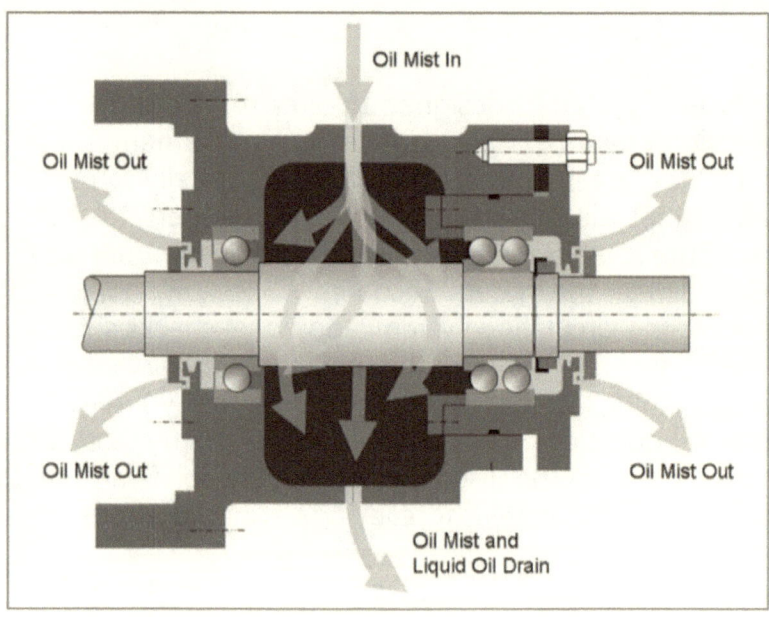

Figure 2.8—Oil Mist Lubrication

Operator's Guide to Rotating Equipment

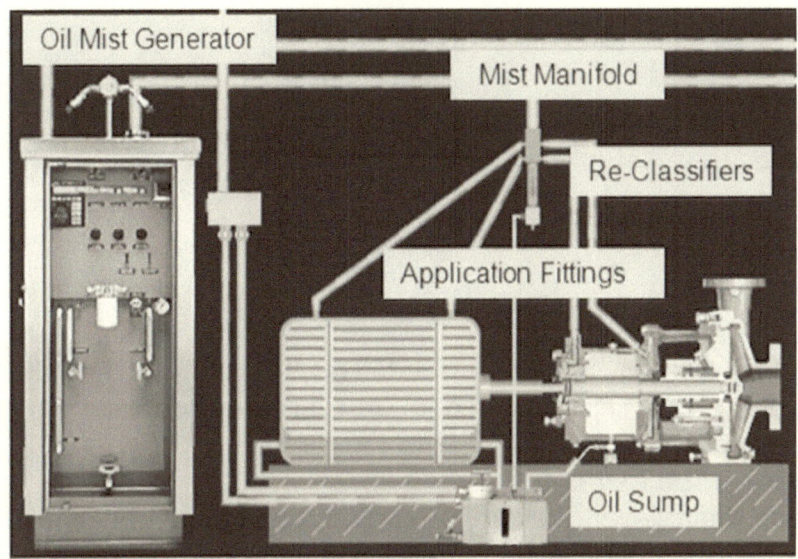

Figure 2.9—Oil Mist Lubrication System

Automatic grease lubricators—These are often used to grease inaccessible machine locations. They have a built-in timing mechanism that can be electronic, chemical, or mechanical (see Figure 2.10). They may or may not put out much pressure, so in some instances they can be prevented from delivering lubricant. It is important for these lubricators to be marked on their reservoirs with a permanent marker line showing the lubricant level and the date that the lubricant was at that level. It is one of the few ways to know that lubricant is being supplied to the intended locations. This technique provides the operator with a visual indication if lubricant is being provided. If the lubricant is being delivered slowly, it should be remarked every few months with a new line and dated to ensure lubricant is being delivered continuously.

Figure 2.10—Auto Grease Lubricator

Lubrication inspections:
Have the following recommendations regarding lubricant properties and oil levels in mind when inspecting lubrication systems:

- Pay close attention to any changes in the lubricant's smell, color, and level. A dramatic change in color may indicate contamination with water (see Figure 2.11). ("The Nose Knows—Using Odor as a Test for Your Oils" in Chapter 6 describes how to use your sense of smell to spot lubrication problems.)
- If the reservoir is small and an oil leak is detected, check the level immediately since not much leakage can be tolerated before the level is dangerously low.
- If the color has changed significantly from your last inspection or from what you know to be normal, find out why.
- Ensure lubricant levels are maintained as recommended. Inadequate oil levels can cause equipment to fail catastrophically in a relatively short period of time.
- If the level is found to be increasing, water may be entering the system in some way. Tip: A simple test for water contamination is to drain some of the lubricant out onto a napkin or paper towel. The oil will be

Operator's Guide to Rotating Equipment

absorbed into the towel but the water beads will stand on the oil-soaked towel.

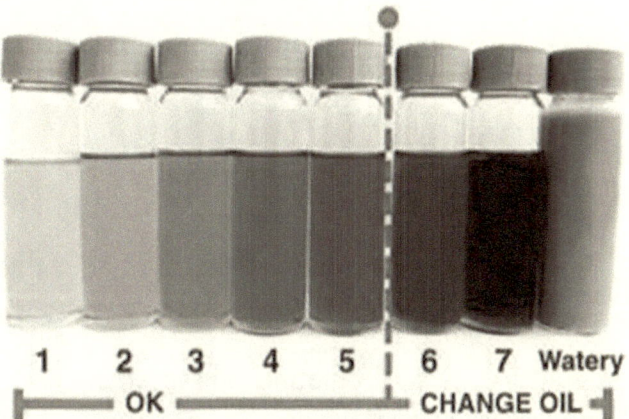

Figure 2.11—How oil appearance can change as it become contaminated. Remember look for changes from newly installed oil.

1. Sample #1 shows ideal oil that is golden and fully transparent.
2. Sample #5 shows a sample that is discolored.
3. Sample #6 shows oil that is no longer transparent and has darkened.
4. Sample #7 is now completely dark due to oxidation.
5. The "Watery" sample (on the far right) is seriously contaminated with water.

Chapter 2
Review Questions

1. List at least 3 critical functions that lubrication provides.

2. _____ is a mixture of a semi-fluid thickening agent and liquid lubrication.

3. List 3 elements of circulating oil systems.

4. List at least 3 ways that lubrication is applied to bearings.

5. List at least 3 things to be looked for during oil inspections.

6. _____ lubrication, also called hydrodynamic lubrication, is where two surfaces are completely separated by a fluid film.

Chapter 2
Answers

1. List at least 3 critical functions that lubrication provides.
   - Reduce friction between the rotating and stationary components
   - Reduce wear
   - Absorb shock
   - Dampen noise
   - Carry heat generated by friction within the bearing
   - Remove heat transmitted down the shaft from the process end of the machine
   - Minimize corrosion
   - Keep contaminates away from the bearing components

2. <u>Grease</u> is a mixture of a semi-fluid thickening agent and liquid lubrication.

3. List 3 elements of circulating oil systems.
   - Reservoir
   - Pump or pumps
   - Filter or filters
   - Heat exchanger

4. List at least 3 ways that lubrication is applied to bearings.
   - Splash
   - Ring
   - Circulating
   - Forced
   - Oil mist
   - Automatic grease lubrication

5. List at least 3 things to be looked for during oil inspections.
   - Oil level
   - Unusual smells
   - Oil appearance
   - Oil color

6. <u>Full fluid</u> lubrication, also called hydrodynamic lubrication, is where two surfaces are completely separated by a fluid film.

Chapter 3
Inspection techniques available to operators and field personnel

Audible inspections

Audible, or sound, inspections are done during routine process rounds. Listen to the equipment and notice differences that may have occurred since the last time you were near this piece of equipment. The listening must be done regularly so changes can be detected. It is necessary to know what good sounds like to be able to tell what is "bad". The following examples of observations may be telling you something:

1. A loud hum on a motor may indicate a soft foot condition. (Soft foot is a condition where a motor foot is out of plane with the others, which leads to a warping of the motor frame.) If it was quiet yesterday and noisy today, you need to seek the reason for the change.

2. Squealing belts can indicate anything from an overload condition to just loose belts. A scraping sound or rhythmic sound can indicate a dragging or rubbing problem.
3. A clicking or regularly recurring sound from inside the gear box may indicate a broken tooth. If there is a sight glass on the oil system, see if metallic particles can be detected floating in the oil or try the paper test. (Described later in this book.)
4. Tapping heard on a reciprocating compressor usually indicates a loose valve. Use a listening device on the covers one after another to find the loose valve. This will ensure that only the loose valve's cover needs to be opened to replace the valve seat gasket and the valve's securing jack screw tightened before the valve seat is damaged. As a temporary "fix" the hold down device for the valve may be tightened to secure the valve but it could still be leaking.
5. A failing or poorly lubricated rolling element (or frictionless) type bearing can often be detected audibly.

Visual inspections

Operator Checking Seal Pot Level

Always be on the lookout for leaks, abnormal fluid levels, burned paint, and vibrating shafts or housings. Here are some specific situations to keep in mind when inspecting:

1. Check for any signs of discolored or burned paint. Paint burns at about 400 to 450 degrees F. If the burned paint is near lubricating oil, it is likely the oil is hot as well. This high temperature indication could mean the condition of the oil is no longer viable as a lubricant.
2. Look at motor fan guards for blockage and check the fins on TEFC motors to ensure cooling can take place.
3. Ensure pressure gauges are installed and working. It is most important to know what the normal pressure is for any piece of operating equipment. Sometimes gauges are marked with the acceptable range. If this is done on the gauge glass, make sure there is another mark to indicate the correct glass position, ensuring that it has not moved and is giving a false reading.
4. Look for signs of over-lubrication, which will cause leaks and shorten equipment life. It is also a potential environmental problem and fire hazard (see Figure 3.1).

Figure 3.1—Over-lubricated Bearing

If equipment is ring oiled, it is generally easy to look at the ring while the equipment is running. This is an excellent check, because if the ring stops turning for any reason, it has the same effect as a loss of oil flow in a forced lubrication system. Improper operation of an oil ring will likely lead to a bearing failure.

Tactile inspections

Tactile inspections are easy to perform and can tell you a lot about how the machine is running. But they are to be carried out only if you are sure you will not be burned by touching the equipment. First, swipe the equipment with a brushing motion of the fingertips. If it is not too hot, gradually slow the motion to determine if it will become uncomfortable to the touch. If you can leave your hand on the equipment the following states of operation may be examined:

1. Lightly touching a piece of equipment with fingertips can provide a subjective evaluation of how smoothly it is running. With practice, manual vibration checks can be a reasonably good method of detecting mechanical changes. A tingling in your fingers when touching the machine indicates a high frequency or fast occurring vibration. It is recommended that vibration checks be done regularly in order to note changes from yesterday's or last week's vibration level.
2. If any of the surfaces are hot, ask yourself the following questions: a) Does it seem hotter than the last time you touched it? b) Is the hot spot localized or is it generally hot? This is excellent data to share with engineering or maintenance.
3. If a seal pot is used, touch the two lines going to the seal gland. If fluid circulation is taking place in these lines the top line should be hotter than the lower. The same can be done for the cooling water lines into the seal pot if it is water cooled. This is a way to know if circulation is taking place. This technique can be used

on any type of heat exchanger to get a gauge for heat transfer taking place.
4. If a pump has multiple filters, such as in an oil or hydraulic system, touching the outlet of the pump and then each filter downstream can help to determine which filter is in service. The filter with the same temperature as the pump discharge is the one in service.
5. A leaking pressure relief valve can be detected by touching the discharge side of the relief valve and the inlet to the relief valve. There should be no flow through the valve unless there is an overpressure. If there is leakage or if the valve is relieving there will be no difference in the temperatures of the two lines. If the temperatures are the same, it is likely that the relief valve is relieving or leaking. This condition could divert needed lubricant flow from the equipment and cause the stand by pump to come on.

Smell

Smell is another means of detecting a change in machinery. Here are examples of how your sense of smell can be used as a monitoring tool:

- If belts are loose, it may be possible to detect this situation not only by the sound but by the smell of the hot rubber as it is removed from the side of the belts.
- Burned oil has a distinct scent that can indicate a potential problem.
- Paint that gets hot enough to discolor gives a distinctive odor. Each scent may indicate a particular problem with the equipment.
- A problem can also be detected if the product processed in the plant gives off a "normal" odor and a distinctly different odor when overheated or when a problem exists.

The advantages of the audible, visual, tactile, and smell-based inspections are that they are:

- Easy to use
- Easy to train someone to do
- Always available
- Inexpensive
- Can be done by anyone

The disadvantages are that they are:

- Subjective (Hot? Very hot? Burning up? What does that mean?)
- Difficult to communicate to work request or someone else
- Difficult to repeat (Different people perceive the same thing differently)
- Unable to be used at the earliest stage of a problem

Tools available to quantify what you have detected:

Audible Inspection Methods

Ultrasonic Gun
An ultrasonic gun is a relatively inexpensive handheld device for listening to noises in the ultrasonic range. It is used for leak detection, steam trap maintenance, and listening to bearing noises.

Stethoscope
This is an inexpensive tool that can pick up every sound that the equipment is making, though that can also be counterproductive. The good news is that it picks up everything and the bad news is that it picks up everything. It must be used regularly to detect changes from the last inspection, or it is likely to be useless in determining if there is a problem or not.

Metal Rod
A steel or aluminum rod can be used to touch the piece of equipment where the suspected noise is located, with the

other end touched to the ear to listen for unusual noises. The end touching the ear should be padded, and the other end should be kept away from rotating shafts. A screwdriver can be used or a tool can be improvised by using a valve wrench or even the edge of a hard hat. The goal is to collect the sound of what is happening inside rotating equipment, listening for potential problems.

Visual Inspection Methods:

Infrared or IR gun

This handheld, noncontact temperature measuring device is an easy tool to use if its limitations are understood. One Important limitation of an IR gun is that the actual sensing area of the device is cone shaped. While the laser pointer indicates only the very center of the cone, the temperature being measured can be quite large depending on how far away from the item being measured the IR gun is. The reading is an average of what the cone sees. The effect of the ever-widening sensing cone is that the further away you are from the item of interest, the larger the surface area that will be averaged into the reading (see Figure 3.1). The laser represents a spot that is the center of the cone that is being measured. Figure 3.1 represents how the averaged reading area gets larger as the measure point is further from the measuring device. The closer to the object being measured the readings are taken the better the data will be.

The first surface that the device encounters is the one that will be measured. If you are trying to measure the temperature of something behind a glass or plastic cover, the temperature of that cover is what will be measured. If you wish to obtain accurate surface temperatures, then the device must be used on flat, dark-colored objects. IR guns return very low readings on shiny surfaces. It is always most through to take a reading on a piece of paper that is at the temperature of the surroundings you wish to take readings. This will represent the "ambient" readings.

For the best temperature measurement results, paint a black spot on the point of interest and use your IR gun as close as possible to the black mark. This method is the only sure way to get useful, repeatable readings. Fig 3.2 indicates the measuring area increases the further from the IR gun you are. For best results hold it as close to the object as you can.

Single point laser

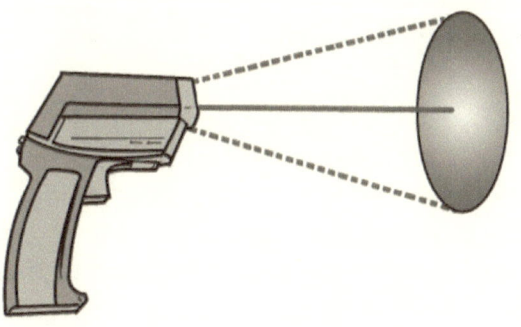

D:S = 6:1 at a 36" distance you are actually measuring a 6" spot not the center of the laser dot.

Figure 3.2—Infrared Temperature Gun

IR Camera
An IR or infrared camera is a device that forms an image using infrared radiation, similar to a common camera that forms an image using visible light (see Figure 3.3). The temperature of objects in the output screen may be depicted in a gray or color scale range, depending on the type of IR camera you are using. A scale indicating what the displayed gray or color scales mean is usually in the output screen as an aid to help the user determine the temperatures of the various points being surveyed. What may appear to not be a problem with the naked eye may be seen very differently with the infrared camera.

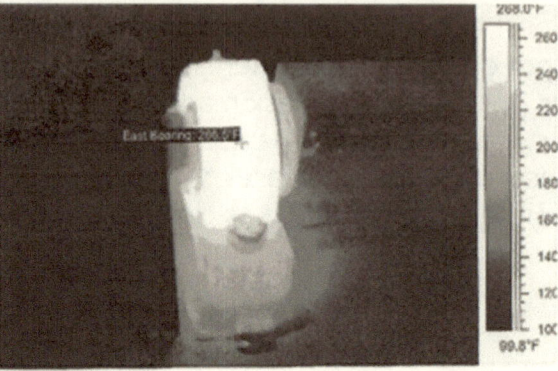

Figure 3.3—IR Camera Picture

Strobe light

This device has a high intensity light that can be controlled to a specific flash rate (see Figure 3.4). When an object is turning at a specific speed and the strobe flash rate is tuned to the same flash rate, the moving object appears to be stopped. This capability allows a good visual inspection to be done while a piece of equipment is turning at running speed. Issues like broken shim packs in couplings, missing keys, or broken fans on electric motors can all be inspected without having to stop the piece of equipment. With a small mismatch of flash rate it will make the object viewed appear to rotate slowly so an inspection of the entire circumference can be inspected. If a spinning object has an unknown RPM by adjusting the strobe through a range the speed can be determined.

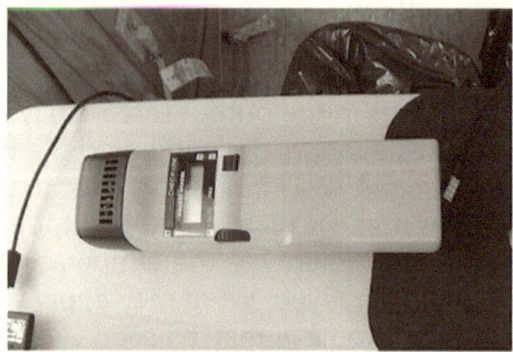

Figure 3.4—Pocket Strobe Light

Inspection Methods Using Vibration and Temperature Measurement Equipment:

Vibration Meter with Accelerometer

A hand held vibration meter (see Figure 3.5) with a mag-based accelerometer can be used to collect overall casing vibration readings as a means of gauging the mechanical condition of a machine. This additional monitoring tool can augment the five senses by providing information that is objective instead of subjective. Everyone that uses the same type of vibration equipment in the same place should get the same value; if 50 people touch the same piece of equipment with their hands in the same place, there are likely to be 50 different interpretations of how much vibration there is.

Figure 3.5—Technician Using Vibration Meter with an Accelerometer

Figure 3.6 illustrates how a vibration trend can be used to identify a problem. In this example, vibration data have been collected and plotted on a weekly basis. The magnitude of vibration is reported in units of velocity (inches per second).) It is easy to see that in week 10, a detectable change in vibration occurred.

Trend plots are useful because they provide visual representations of the measured parameter (e.g. vibration, temperature, pressure, etc.) over time, which can be helpful in troubleshooting and the decision-making processes. Suppose a step change in vibration occurred at the same time as a change in the process; this fact may point to a correlation between the two events that should be investigated. Always keep in mind that a gradually increasing trend plot may indicate either a deteriorating internal or external component and should be investigated once detected. Most measurements are in velocity and as a general rule under .2 inches per second (IPS) is acceptable. The range of .3 IPS and higher is beginning to get rough. Generally the lower the number the smoother the machine is operating. There must be a trend as below to know what is acceptable and when it is not.

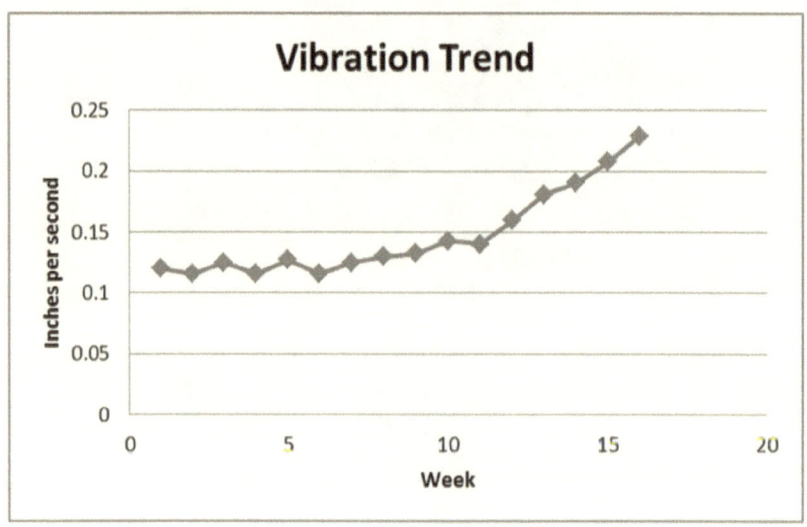

Figure 3.6—Vibration Trend Example

Temperature measurement equipment

A comparison of subjective versus objective evaluations also holds for temperatures measured by touch versus a calibrated temperature device. If a quantitative analysis is required, a contact thermometer is an inexpensive and accurate tool to measure temperatures. It is simple to use and will give repeatable results no matter the color of the surface being measured as long as it is placed in the same place each measurement.

Figure 3.7 shows weekly temperature trends of two different bearing temperature measurements. It is easy to see that the two bearings seem to track together. However, when a bearing problem surfaces in bearing #2, its temperature begins to increase relative to bearing #1 after week 10. This simple visual trending method can help to identify developing machine issues.

These techniques visually explain the need to know what it was yesterday, last week, last month, etc.

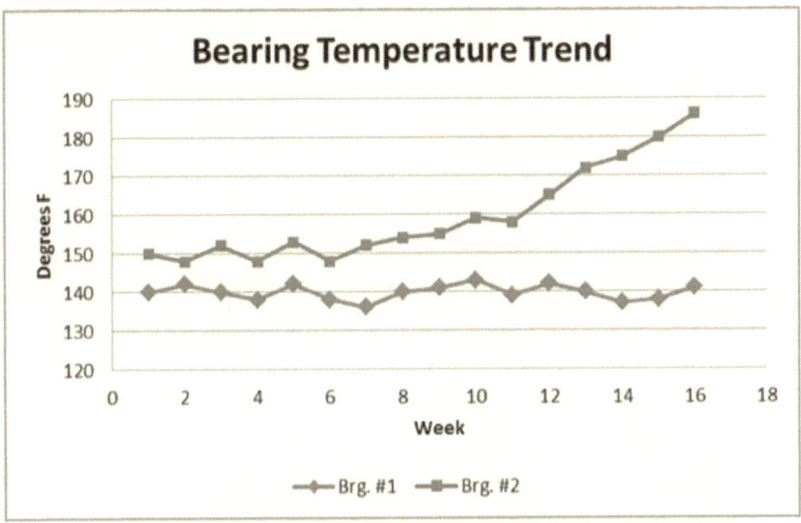

Figure 3.7 Temperature Trend

Putting all the data needed in one place in general term the following is suggested?

- Oil temperatures should generally be above 100 F before starting equipment.
- Temperatures above 180F are usually indicative of a problem.

- Vibration levels below 0.2 IPS (inches per second), are generally acceptable. Above 0.3 IPS vibration is starting to get rough and above .5 IPS is very rough.

Remember these are general guidelines. Always compare current operating values with those when the machine was operating normally. If you are unsure involve others to help make a determination on the next course of action.

Chapter 3
Review Questions

1. What senses are available to an operator to monitor machinery?

2. List three ways that smell can be used to detect a machinery problem.

3. List three ways that sound can be used to detect a machinery problem.

4. List three tools that can be used to monitor temperature.

5. List three things to look for during visual inspections of machinery.

Chapter 3
Answers

1. What senses are available to an operator to monitor machinery?
   - Hearing
   - Smell
   - Touch
   - Sight

2. List three ways that smell can be used to detect a machinery problem.
   - Slipping belts
   - Burning paint
   - Degrading oil

3. List three ways that sound can be used to detect a machinery problem.
   - Electric motor problems
   - Broken tooth on a gear box
   - Loose valve in a reciprocating compressor
   - Failing or poorly lubricated rolling element bearing

4. List three tools that can be used to monitor temperature.
   - Contact thermometer
   - IR gun
   - IR camera

5. List three things to look for during visual inspections of machinery.
   - Notice if there is any unusual vibration
   - Notice if there are any hot surfaces
   - Check for leaking relief valves
   - Check temperatures going into and out of seal pots to make sure there is a differential

Chapter 4
How to Inspect Process Machinery

Operator Checking Seal Pot Level

The inspection methods described in this section are primarily limited to those used on centrifugal and positive displacement pumps, gear boxes, motors, heat exchanger and turbines. However, with some modifications, the inspection methods can be applied to all types of equipment, such as fans, compressors, extruders, turbines, motors, conveyers, elevators, etc.

You will notice a common pattern after studying the best inspection practices discussed below. In all cases, you are asked to look, listen, and feel. These simple methods can help locate basic machinery problems early. They will help you to determine when a piece of equipment is needed to quantify the magnitude of the problem and get help. With practice, these methods should become second nature.

Recommended Pump Inspection

1. **LOOK** at the pump as you walk up to it.
   - Is it shaking?
   - Is it smoking?
   - Is the discharge pressure different today than yesterday? Is it steady or wiggling? A wiggling needle on a pressure gauge may be due to cavitation.
   - Are all of the anchor bolts in place? Are they tight?
   - Are there indications of leakage of fluids of any kind? (see Figure 4.1)
   - Have parts of the pump come loose?
   - Look at the motor amps. Pump capacity problems due to flow restrictions or internal wear can cause the motor to pull less or more than the normal amp load, depending on the pump design and the nature of the problem. This means it is important to regularly monitor amp loading and note any deviations, high or low, from normal values.
   - Is the seal leaking? If this is a packed pump is the leakage rate excessive? (Note: Mechanical seals should not have any visible leakage, while pump packing requires visible leakage for cooling and lubrication. A

packing leakage rate of about one drop per minute should be considered normal.)
- If the seal uses a seal pot, check to see if the sight glass is showing a correct fluid level? Is the line to the top of the seal gland connected to the upper fitting on the seal pot?
- Is there coupling spacer dust on the foundation? Pieces of shim pack?
- Is the oil level in the bearing housing correct (see Figure 4.2)? Discolored?
- Is oil pressure correct for the pump?
- Is the ΔP, pronounced delta P, (upstream pressure minus downstream pressure), on the oil filter gauge low? Is delta "p" on the oil filter high? If the ΔP on the oil filter gauge is high, can you explain why?
- Is paint burned off in new places? Why? (Remember that paint discolors at around 400 to 450 degrees F.)
- If the pump is equipped with a permanent vibration monitor, check it on a regular basis, looking for any dramatic changes between inspections.

Figure 4.1—An obvious indication of a leak.

Operator's Guide to Rotating Equipment

Figure 4.2—An oil level sight glass showing the full oil level.

2. **LISTEN** to the pump.
   - Is it noisy? Can you detect a bearing noise? Can you hear any cavitation noise? The louder the cavitation noise the greater the potential for internal damage over time.
   - Does it sound different today than yesterday? Is the noise from the motor or the pump?
   - Is the noise constant or changing? (This may be an indication of a control valve opening and closing.)
   - Does it sound like gravel inside of the pump casing? (This could possibly be a sign of cavitation.)
   - Are there any steam, air, or gas leaks in or near the pump?
   - Are the drive belts squealing? Are there any visual indications of loose belts?

3. **FEEL**—Touch the pump with finger tips.
   - Is it warm, hot, cold?
   - Is it different than it was yesterday? How and why is it different than yesterday?
   - Is it shaking more than yesterday? Is it too much?
   - Is the auxiliary oil pump running? Why?
   - Is the oil relief valve dumping oil? Why?
   - Is the pump vibrating? Is it the same as yesterday?

- Is the bearing housing hot? Touch the seal flush lines. Can you tell if there is a difference in temperature? No difference can indicate a loss of flow or a reduction of heat exchange?

Recommended Electric Motor Mechanical Inspection

Operator checking electric motor vibration

1. **LOOK** at the motor.
   - Is it shaking?
   - Is it smoking or are sparks flying?
   - Is anything loose, shaking, or vibrating on the motor?
   - Is the flex conduit in good condition?
   - Has the dust cap come off the bearing on the coupling end of the motor?
   - Is the fan on the TEFC (totally enclosed fan cooled) motor turning?
   - Are the air filters or fins on a TEFC motor clear so air can circulate (see Figure 4.3 and 4.4)?
   - Is there any burned paint? This would be a clear indication of an abnormally high motor surface temperature. If yes, why and where?

Figure 4.3—Electric motor with debris beginning to block the air inlet screen. This could lead to overheating and the eventual failure of the internal windings.

Figure 4.4—A TEFC (totally enclosed fan cooled) electric motor with debris resting between the cooling fins. This reduces the motor's ability to expel internal heat, which could lead to overheating and the eventual failure of the internal windings.

2. **LISTEN** to the motor.
   - Does it sound different today than yesterday?
   - Is the noise the motor or the driven piece of equipment?
   - Is the noise constant or a rhythmic hum?

- Does the noise seem to coming from the bearings, the fan, or some other location on the motor?
- Are belts slipping on the drive end of the motor?

3. **FEEL**—Touch the motor.
   - Is it different than it was yesterday? How and why is it different than yesterday? Is it warm, hot, cold?
   - Is it vibrating?
   - Is the fan turning and putting out air?
   - Is the cooling system for the motor and/or lubrication system functioning properly based on your experience?

Recommended Gear Box Inspections

Figure 4.5 Industrial Gearbox

1. **LOOK** at the gearbox as you walk up.
   - Is it shaking?
   - Smoking?
   - Are all of the anchor bolts in place? Are they tight?
   - Is there indication of leakage of fluids of any kind?
   - Have vibrating parts come loose? (Check oil coolers, piping, base plate, bearing caps, etc.)
   - Is there water in the oil?
   - Is the auxiliary pump running? Why?
   - Is the oil pressure correct?

- Is the oil cool enough? If the oil supply line has any type of temperature indicator, see if the temperature is in the normal range; if there is no indicator, touch the supply line to determine if it seems to be consistent with past inspections.
- Is the oil level correct in the sump?
- Are there coupling pieces on the pedestal under the coupling guard?
- Is the ΔP on the oil filter gauge high (see Figure 4.6)? Why?
- Check the vibration readings log if continuously monitored.

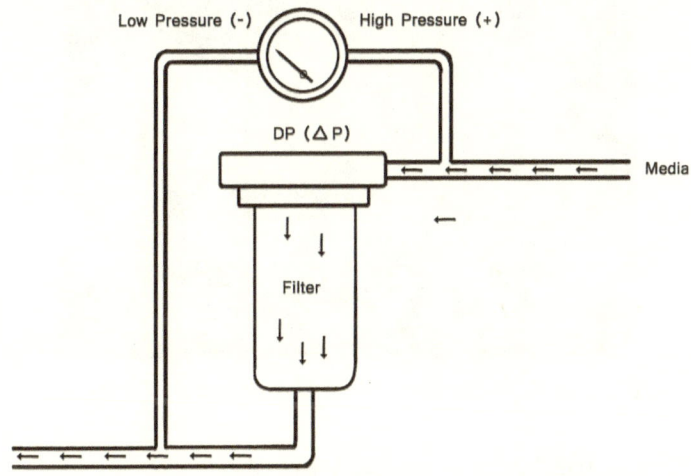

Figure 4.6—A local differential pressure meter, if provided, can be used to indicate the pressure difference between high (upstream) pressure and low (downstream) pressure across an oil filter. This is an excellent means of monitoring the condition of the filter.

2. **LISTEN** to the gear box.
   - Is it noisy?
   - Does it sound different today than yesterday?
   - Is the noise constant or changing?
   - Are there steam, air, or gas leaks in or near the gear box that could contaminate the lubrication?

3. **FEEL**—Touch the bearing housings with your fingertips.
   - Is it different than it was yesterday? How and why is it different?
   - Are they excessively hot?
   - Is it shaking more than yesterday? Is it too much?
   - Is the oil pressure relief valve bypassing oil? If so why?
   - If there is an oil cooler is heat being exchanged? Touch the inlet and outlines to ensure heat is being removed.

Recommended Steam Turbine Inspection

Figure 4.7—General Purpose Steam Turbine

1. **LOOK** at the turbine as you walk up.
   - Is it shaking?
   - Smoking? **Any** signs of smoke, which may be an indication of an oil leak, should be acted on immediately due to fire potential it represents.
   - Are all of the anchor bolts in place? Are they tight?
   - Are there any indications of fluid leakage?
   - Are there vibrating parts on the turbine that have come loose? (Check coolers, piping, baseplate, bearing caps, etc.)
   - Is excessive steam leaking out of the glands that seal the shaft to the casing? (Note: Normally, there should be <u>no</u> visible leakage exiting the turbine seal glands.

Any significant gland leak has the potential of injecting steam into the bearing housings and contaminating the oil.
- Is there water in the oil? (Check level, is it rising?)
- Is the governor hunting? (This wears out valve packing and linkage.)
- Are the steam traps near the turbine working (see Figure 4.8)?
- Is the auxiliary pump running? Why?
- Is the oil pressure correct?
- Is the oil cool enough, based on your experience with this steam turbine? Is the cooler working?
- Look at the vibration readings for the turbine; are they steady and low? Is the oil level correct in the sump? Is the oil level correct in the bearing boxes? Is the oil level rising or falling?
- Are the ring oilers turning or hung up?
- Is there steam leaking out of the stem of the control valve?
- If installed, is the bearing housing air purge turned on to prevent steam from contaminating the oil?
- Are there coupling pieces on the pedestal under the coupling guard?
- Is the trip mechanism resting on its knife edge?
- Is the ΔP on the oil filter gauge high?
- Look at piping support springs to insure that blocks were not left in after maintenance, especially if a hydrostatic, i.e. "hydro", test was performed on the piping system during any construction work.
- Look at the coupling area and see if there are shims from the spacer or dust if an elastomeric type of coupling is used.

Figure 4.8—Condensate or free water in a steam turbine's inlet will lead to the rapid erosion of the steam path components, which is why steam traps are always installed in steam turbine inlet piping. A steam trap is used to remove condensate and non-condensable gases from steam piping, with a negligible consumption or loss of live steam.

2. **LISTEN** to the turbine.
   - Is it noisy? Steam leaking?
   - Does it sound different today than yesterday?
   - Is the noise constant or changing? Is the governor steady or hunting?

3. **FEEL**—Touch the turbine bearing housings with your fingertips.
   - Are they excessively hot?
   - Is the temperature different than it was yesterday? How and why is it different than yesterday?
   - Is it shaking more than yesterday? Is it too much?
   - Is the oil pressure relief valve bypassing oil? Touch the inlet to the valve and the outlet from the valve. Under normal conditions, the outlet line should be noticeably cooler that the inlet line. If the relief valve is leaking, the outlet line will feel warm.
   - If it is bypassing oil, try to determine if this is normal or not.
   - Check oil cooler to ensure it is removing heat from the oil.

Recommended Heat Exchanger Inspection

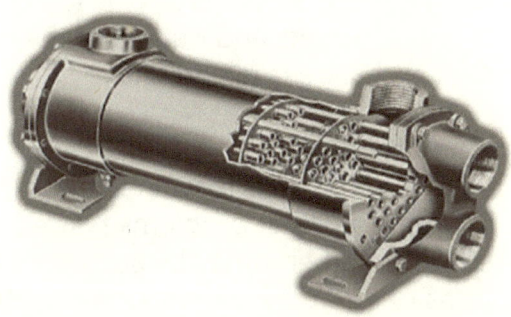

Figure 4.9—Heat Exchanger

1. **LOOK** at the heat exchanger as you walk up.
   - Is it shaking?
   - Are all of the anchor bolts in place? Are they tight?
   - Are there any indications of leakage of fluids of any kind?
   - Is the differential pressure correct?
   - Is heat being exchanged in the cooler? Touch the inlet and outlet and insure there is a difference in temperatures.
   - Is the delta temperature for the exchanger normal?
   - If the exchanger operates hot, is the mechanism that allows controlled expansion clean, rust free and functioning? Are slides, flex legs, or bolts left loose on one end to allow movement of one end of the exchanger?

2. **LISTEN** to the exchanger
   - Is it noisy?
   - Does it sound different today than yesterday?
   - Is the noise constant or changing?
   - Is there the sound of gas or boiling going on inside?

3. **FEEL**—Touch the exchanger your fingertips.
   - Is it excessively hot?

- Is it different than it was yesterday? How and why is it different than yesterday?
- Is it shaking more than yesterday? Is it too much?
- Touch or test the inlets and outlets of the exchanger to see if an exchange is taking place.

General Machine Start-ups

During the start-up of a process machine, it is prudent to be extra vigilant in order to detect problems related to how a machine is lined-up, whether it may be subjected to adverse process conditions or equipment deterioration, etc. Extra attentiveness should be given to machines that have recently been repaired or idled for extended periods of time.

Keep in mind that the instructions provided here are very general and should be performed on the driver and the driven equipment before and during start-ups. More specific start-up procedures may be required for more complex or critical machines. Furthermore, before attempting to start-up a critical machine, verify the locations of all key field instruments, such as flow meters, pressure gauges, and temperature gauges and know what are considered normal indications for all these devices (see Figure 4.10).

Figure 4.10—Always locate all key field instruments before starting rotating equipment. The figures shown here are of an oil pressure gauge (upper left), differential pressure gauge (lower left), level glass on a reservoir (center), and a rotameter (right)

Inspection:

- Check and start all auxiliary system, including lubrication, seal, and cooling systems, depending on what your specific machine has installed
- Ensure there are adequate levels in all components that have liquids, including sumps, lubricators, greasers, and barrier or buffer systems, etc.
- Look at the condition of the lubricant. Color changes, especially if they happen rapidly or are not normal for this piece of equipment, should be investigated prior to starting.
- If this equipment operates hot, allowances must be made for warm up to allow all parts to come to normal operating temperatures.
- If there is a device such as a guided slide or flex plate, as on steam turbines, it must be free to allow movement or flexing as temperatures approach normal from ambient. The same is true for equipment operating at extreme cold temperatures.
- On motors, ensure ventilation openings are clear of obstructions that could restrict airflow.
- Ensure the area around the equipment is clean and free of hazards.
- For motors, verify there are no loose conduit or cable connections or broken conduit. Ensure all foundation bolts are tight.
- If there are site-specific or manufacturer's specific instructions, they must be incorporated in your operating procedures.
- Start the equipment.
- If the unit is motor driven and does not start the first time, look for all the reasons why it didn't start. These could include the activation of one or more permissives that have been installed to protect the motor from starting before conditions are safe. (A permissive is a particular type of safety interlock designed to prevent actions from taking place until

pre-defined criteria have been satisfied. A permissive may prevent machine start-ups if there is a condition of low oil pressure, high suction drum level, high winding temperatures, etc.) Most large motors will allow only 3 starts then they lock out for hours before they can be started again.

Machine Shut-downs:

Operating procedures should also be developed and followed for equipment shutdowns. Some auxiliary systems, such as lubrication to protect bearings from overheating and seal oil systems to maintain pressure differentials, may need to be kept running. The manufacturer is an excellent source for shutdown guidelines.

Keep in mind that the instructions provided here are very general and should be performed on the driver and the driven equipment after machine shut-downs. More specific shut-down procedures may be required for more complex or critical machines.

Inspection:

- Prior to shut down, look at the condition of the lubricant while it is still circulating. Color changes, especially if they happen rapidly or are not normal for this piece of equipment, should be investigated.
- If this equipment operates hot, allowances must be made for cool down to allow all parts to come to ambient temperature.
- Ensure there are adequate lubricant levels in all areas that require it, i.e. sumps, lubricators, greasers, etc.
- If there are site-specific or manufacturer's specifications, they must be followed.
- Stop the equipment.
- Stop the auxiliary systems, which include lubrication, seal systems, and cooling. If rotors are extremely hot (Above 250°F) allow the lubricating system to circulate

to cool the shaft and bearings. This is especially necessary if the bearings are made of babbited material.
- If the equipment has a cooler to regulate temperatures it is ideal to have a method of back flushing the cooler and this should be done at each opportunity such as shutdown or equipment swaps.
- Ensure ventilation openings on motors are clear of obstructions that could restrict airflow.
- Verify there are no loose conduit or cable connections or broken conduit on motors.
- Ensure all foundation bolts are tight.
- Perform a visual inspection for leaks after shutdown.

For turbines make sure steam traps are working to ensure the turbine does not fill with water during shut down.

Chapter 4
Review Questions

1. What are the three senses you can utilize during a machinery walk-though inspection?

2. List three things to look for during a pump inspection.

3. List three things to listen for during an electric motor inspection.

4. List three things to feel for during a steam turbine inspection.

5. The one secret to detecting problems early is to know what _____ is.

Chapter 4
Answers

1. What are the three senses you can utilize during a machinery walk-though inspection?
   - Sight
   - Hearing
   - Touch

2. List three things to look for during a pump inspection.
   - Is it shaking?
   - Smoking?
   - Is the discharge pressure different today than yesterday?
   - Are all of the anchor bolts in place? Are they tight?
   - Are there any indications of leakage of fluids of any kind?
   - Are there vibrating parts on the pump that have come loose?
   - Are the motor amps normal? Pump capacity problems do not cause the motor to pull excessive power or higher amps.
   - Is the seal leaking?
   - Is there proper level in the seal pot if used?
   - Is there coupling spacer dust on the foundation? Pieces of shim pack?
   - Is the discharge pressure gauge steady? (If not there may be cavitation)
   - Is the oil level in the bearing housing correct? Discolored?
   - Is oil pressure correct for the pump?
   - Is the ΔP indication on the oil filter gauge low? High? Why?
   - Is paint burned off in new places? Why? (Paint discolors around 400° to 450° F.)
   - Look at the vibration monitor on a regular basis.

3. List three things to listen for during an electric motor inspection
   - Is it noisy? Bearings? Fan?
   - Does it sound different today than yesterday?
   - Is the noise the motor or the driven piece of equipment?
   - Is the noise constant or a rhythmic hum?
   - Are the belts slipping on the drive end of the motor?

4. List three things to feel for during a steam turbine inspection.
   - Are they excessively hot?
   - Is it different than it was yesterday? How and why is it different than yesterday?
   - Is it shaking more than yesterday? Is it too much?
   - Is the oil pressure relief valve bypassing oil? If so why?
   - Check oil cooler to insure it is removing heat from the oil.

5. The one secret to detecting problems early is to know what <u>normal</u> is.

Chapter 5
An Introduction to Compressor Operations

Figure 5.1—Centrifugal compressor driven by a gas turbine

Compression Basics

Gases are fluids that take the shape of their container and are highly compressible when compared to liquids. Gases can be either pure or a composition of different gases with varying molecular weights. Given two equal volumes of gas at identical pressures and temperatures, the volume containing gases with a higher average molecular weight will have a higher density, which means it will have a greater mass per unit volume. We must know the volume, pressure, temperature, and composition of a gas to fully define its nature and condition.

A compressor is a fluid handling machine that takes in gas at a lower density and pressure via a suction nozzle and compresses it, resulting in the gas having a higher density and pressure in the discharge nozzle. Figure 5.2 demonstrates how compression works: It begins with a given starting volume of gas, as shown on the left in Figure 5.2. The gas is squeezed into a smaller volume, as shown on the right. Notice that the starting volume of the gas has been reduced, while the number of molecules inside the container remains the same. A smaller volume with the same number of molecules means that density has increased. Compressors like centrifugal pumps have larger inlets than discharges making it easy to determine which is which.

The act of compression requires power, provided by drivers such as motors, steam turbines, and gas turbines, to push the molecules into a smaller volume. This, in turn, forces them closer together and pressure to increase. It is the higher pressure that pushes gas flow out of the compressor's discharge nozzle into the process. In addition to an increase in pressure, the act of compression causes the gas to heat up. The greater the level of compression, the higher the temperature increase you can expect.

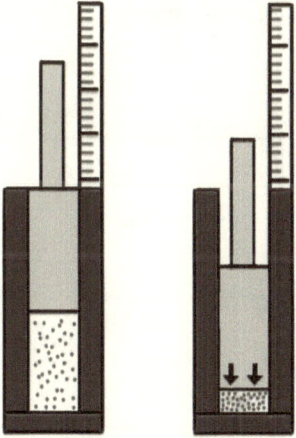

Figure 5.2—During the gas compression process, a volume of gas is decreased in order to increase its pressure as shown on the right.

Defining Gas Flow

One way that compressors are different than pumps is the way flow rates are expressed. Because liquids are essentially incompressible, there is no need to be concerned about the effect of pressure on the flow volume. 100 gallons per minute at 10 psi is essentially 100 gallons per minute at 500 psi. The same cannot be said about gases. Gases are compressible and the density of a gas is highly dependent on its pressure and temperature. Gas flow is expressed in one of two ways: 1) standard cubic feet per unit time or 2) actual cubic feet per unit time.

Standard cubic feet per minute (SCFM) is the flow-rate of a gas corrected to standard temperature and pressure. Expressing gas flow in terms of standard conditions is handy because the value is independent of temperature and pressure. Standard conditions are 14.7 psi (absolute) and 60° F. If we say we have 100 standard cubic feet of a gas it means that at 14.7 psia and 60° F we would expect to have 100 cubic feet of gas. In contrast, the term "actual cubic feet per minute" (ACFM) is used to express the volume of gas flowing anywhere in a system. If a system were moving a gas at exactly standard conditions, then ACFM would equal SCFM.

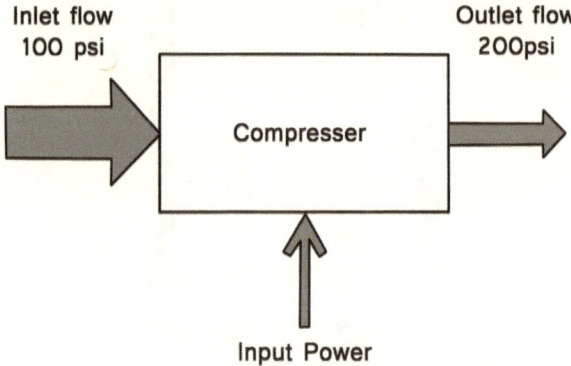

Figure 5.3—Compressor Schematic

Figure 5.3 shows a gas compressor changing a gas stream from 100 psi to 200 psi. The larger input arrow on the left is meant to show that the actual cubic feet per minute is larger than the actual cubic feet per minute at the outlet conditions shown by the smaller arrow on the right. Even though the actual cubic feet per minute value is smaller on the right, the standard cubic feet per minute is the same at the inlet and outlet conditions. The lesson to remember here is that you need to understand which measurement units of the flow you are dealing with when examining compressor performance.

Compressor Types
Compressors come in many different designs in order to handle a wide variety of process applications. The two most common compressor designs are *positive displacement* and *dynamic* (see Figure 5.4). Positive displacement compressors include reciprocating (see Figure 5.9) and rotary units (see Figure 5.5). Positive displacement compressors work by continually forcing gas into a smaller and smaller volume, using either a piston or tight-fitting rotors, and then expelling the reduced volume of gas into a discharge passageway.

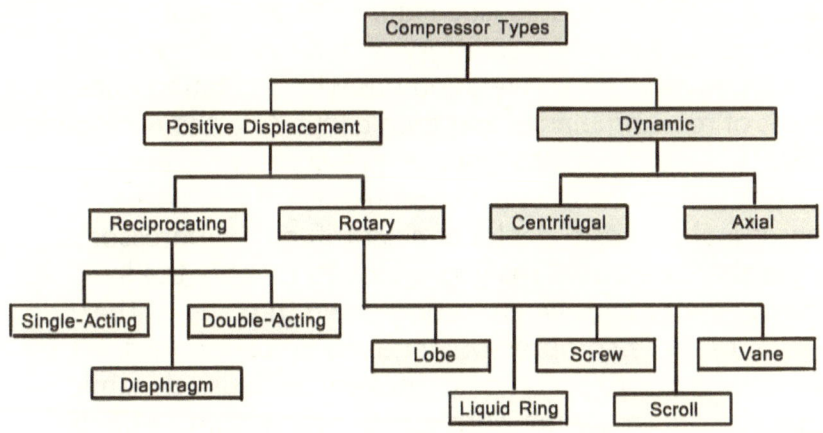

Figure 5.4—Types of Compressors

Figure 5.5 shows a cutaway view of a rotary screw compressor composed of two tightly fitting rotors inside a pressure containing housing. Positive displacement compressors generally deliver smaller gas volumes and much higher differential pressures than centrifugal or axial flow compressors. Piston compressors have a potential drawback of discharging the gas flow in pulses, which some processes may not be able to tolerate.

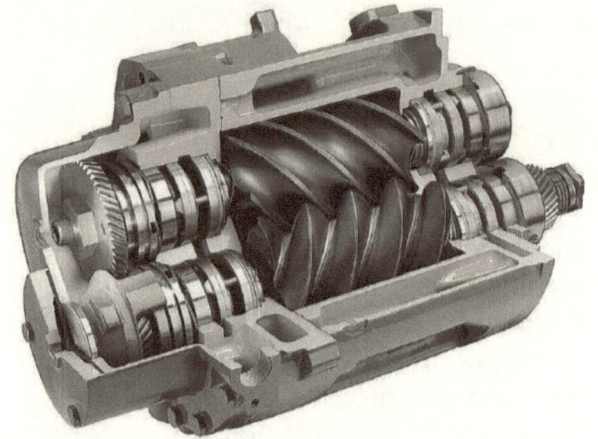

Figure 5.5—Rotary Screw Compressor

Centrifugal compressors (see Figure 5.6) and *axial* compressors fall in the category of dynamic compressors. This type works by accelerating gas with a rotating impeller or set of rotating blades and then converting the exiting high velocity gas stream into a higher pressure by decelerating the gas stream for the next compression stage. Like centrifugal pumps the inlet of a centrifugal compressor is larger than the outlet making it easy to identify the inlet and outline nozzles when tracing down the piping. The output of these compressors tends to be continuous, with very low pressure pulsations. However, because dynamic compressors create pressure due to aerodynamic effects, they are susceptible to a flow condition called surging, which is characterized by violent, periodic flow reversals followed by the reestablishment of flow from suction to discharge. This is a serious operating condition that should be avoided. Surging can be detected either by a characteristic sound or, if equipped with proximity thrust probes, by the periodic back and forth axial movement of the rotor.

Centrifugal and reciprocating compressors are the type most often found in most petrochemical processes. Centrifugal compressors are used whenever high gas flows must be delivered at low to moderate differential pressures. Reciprocating compressors are used whenever low to medium gas flows must be delivered at high differential pressures.

For extremely high delivery pressure (up to 50,000 psi) applications, special reciprocating compressors, called hyper-compressors, are utilized. This chapter will cover only conventional reciprocating compressors.

Multi-staging
Because the act of compression heats up the gas, there is a limit to how gas pressure much can be raised across any given compressor stage. Higher and higher pressures can be achieved by placing multiple stages of compression in series, that is, having the discharge of one stage of compression feed the intake of the next stage and so on (see Figure 5.6). "Multi-staging" is the term used whenever multiple stages are used to meet the pressure requirements of a given application. Both reciprocating compressors and centrifugal compressors can be configured in multistage arrangements. If the temperature from one compression stage is predicted to get excessive, then inter-stage cooling is required to drop the gas temperature to a safe level before entering the next stage. Problems with inter-stage cooling systems can result in loss of flow, pressure capability, and in surging.

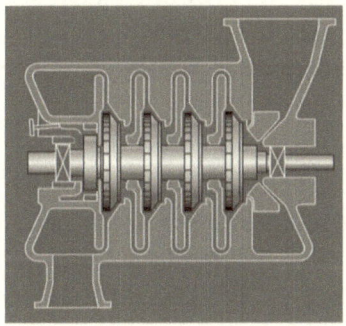

Figure 5.6—Multistage Centrifugal Compressor

Julien LeBleu, Jr. and Robert Perez

Key Reliability Indicators

Some *increase in the gas temperature* is normal and expected. Many compressors have inter-coolers or after-coolers to handle this expected increase in temperature. However, whenever internal compressor issues, such as internal recycling, fouling, etc., are present, gas temperatures can rise well above what is expected. For this reason, compressor temperatures should constantly be closely monitored as a means of detecting internal problems. If high temperature conditions are not detected early, they can lead to the degradation of internal non-metallic elements, which in turn can escalate into a major failure. These high temperatures can also lead to compressor surges. It is best to set a temperature alarm on discharge of every compressor stage that is somewhat below the shutdown level. If the alarm sounds, you will have time to investigate the problem before the compressor either shuts down or you have to shut it down. Early intervention my prevent loss of run time productivity. This is another case where knowing what "normal" is and knowing what the temperature was yesterday, last week, and last month is critical to detecting problems early.

Vibration is another major concern with and excellent indicator of mechanical condition of compressors. Vibration can be caused by imbalance, flow issues, foundation problems, internal looseness, etc. Make sure to monitor piping and skid vibration around reciprocating compressors as well the working components. When reporting a vibration problem, note any special conditions present when the abnormality occurs, such as the rotational speed or process conditions. These notations can be very helpful during troubleshooting later.

Centrifugal Compressors

Centrifugal compressors are a class of gas compressors that operates similarly to centrifugal pumps. The gas enters the eye of an impeller and is accelerated outwardly toward a stationary diffuser that decelerates the gas stream and creates pressure. Typically, multiple stages of impeller/ diffuser sets are required to generate the required differential pressure (see Figure 5.7). The key components that make up centrifugal compressors are the rotor, rotor support bearings, end seals, diffusers between impeller stages, and a pressure casing. The rotor spins inside a pressure containing casing that houses the diffusers, which are designed to convert high velocity flow off of each impeller into higher and higher pressures, until flow reaches the discharge nozzle. Centrifugal compressor performance curves are similar to those of centrifugal pumps. They show that the pressure rise across a compressor increases as speed increases and drops as the flow through the compressor increases (see Figure 5.8).

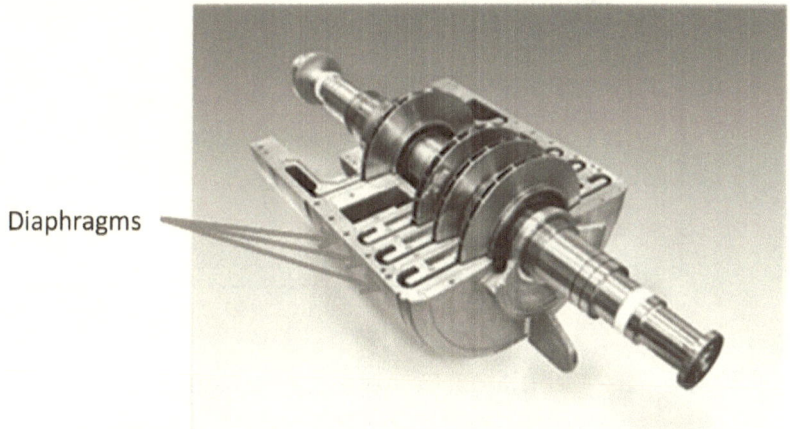

Diaphragms

Figure 5.7—A partially disassembled multistage centrifugal compressor showing the rotor, casing, and flow diaphragms. Diaphragms guide the gas flow from one compression stage to the next.

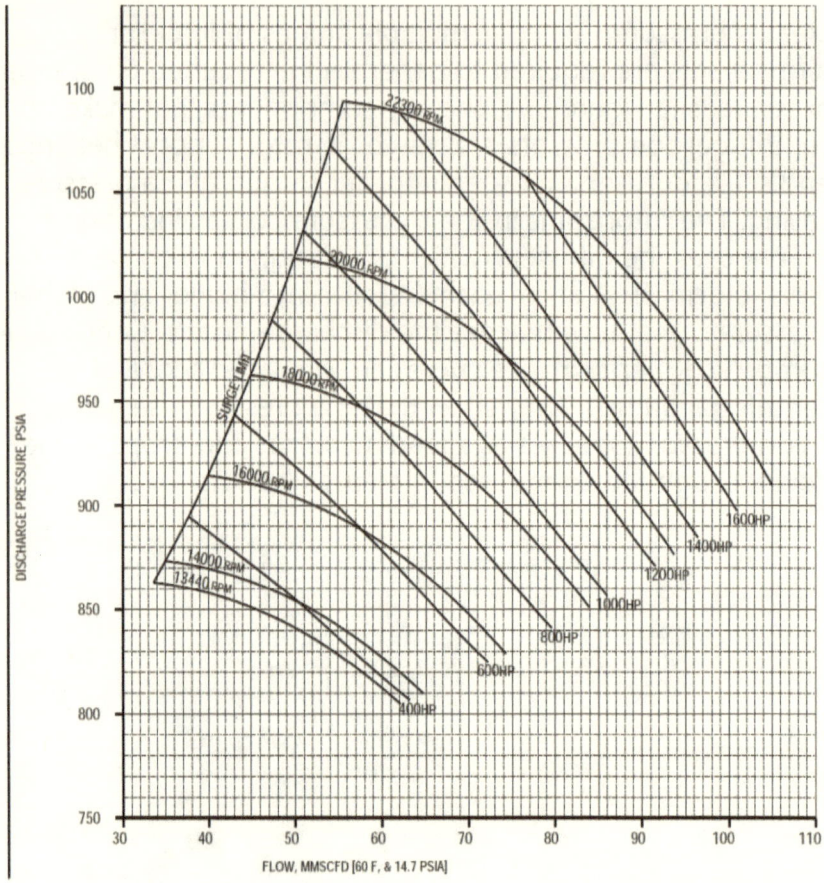

Figure 5.8—Centrifugal Compressor Performance Curve

Troubleshooting Tips:
Here are some useful relationships to remember when dealing with centrifugal compressors:

- If the pressure on the discharge pressure gauge increases, the flow is likely decreasing. Conversely, if the pressure on the discharge pressure gauge decreases, the flow is likely increasing if the suction pressure remains the same.
- If the compressor speed increases, the flow and discharge pressure will likely increase. The horsepower required by a compressor increases at higher

speeds, which should be reflected by increasing amps or kilowatts if the compressor is motor driven. Conversely, if the compressor speed decreases, the flow and discharge pressure will likely decrease as with a turbine driven compressor. The horsepower required by the compressor at lower speed will decrease, which should be reflected by decreasing amps or kilowatts on motor driven compressors or decreasing steam input to the turbine.
- During start up and shutdown or upset conditions, the gas composition can change. If the molecular weight of the gas being compressed increases, the discharge pressure will increase and the required horsepower will also increase. Conversely, if the molecular weight of the gas being compressed decreases, the discharge pressure will decrease and the required horsepower will decrease.
- If the suction pressure increases, the discharge pressure will increase and the required horsepower will also increase. Conversely, if the suction pressure decreases, the discharge pressure will decrease and the required horsepower will decrease. All else being equal, a compressor is a differential pressure machine. Whatever goes in at one pressure will come out at a designed multiple pressure increase. If you raise the inlet pressure, the discharge pressure will also increase by the multiplier of the increase in the inlet pressure.
- Monitor the compressor balance line for temperature differences over time. A significant change can mean a balance piston seal problems, which can result in damaged thrust bearings.

These relationships will help to troubleshoot a problematic centrifugal compressor. When reporting a problem with a compressor, always list the symptoms that caught your attention, such as a low flow, high amps, leaking seals, vibration, not enough pressure, etc.

Centrifugal Compressor Start-ups:
Every compressor is unique in its design and construction and therefore should have its own customized procedure to ensure that all necessary steps are followed during start-ups. Only approved procedures that follow the manufacturer's recommendations combined with operating experience on *this* compressor in *this* service should be used—attention to detail is paramount. If the compressor is motor driven, care must be taken to avoid multiple startup attempts. Generally, if the compressor does not start by the second attempt there is no point in making a third attempt because large motors will generally lock out and there will be a significant cool down time before it can be attempted again. A check should be made of all the permissives after the first attempt to see why the compressor did not start. (A permissive is a particular type of safety interlock used to prevent actions from taking place until pre-defined criteria have been satisfied. For example, some permissives prevent machine start-ups whenever there is a condition of low oil pressure, high suction drum level, high winding temperatures, etc. If these permissives are not cleared before more attempts to start, the motor will likely be locked out for hours before you can try to start the compressor again.)

Once full load conditions (i.e.—speed, temperature, pressure, and flow) are reached, use the "Centrifugal Compressor Checklist" below to monitor machine condition and performance. Continue to monitor compressor conditions for at least an hour to ensure everything is normal. If unresolved issues are encountered during this start-up phase, either call for help in order to get a second opinion or shutdown until additional technical assistance arrives to help you understand what is going on. If a problem is detected, be sure to capture actual data while the problem is present.

Centrifugal Compressor Checklist

- **Net flow through the compressor.** Is the flow too low or too high? If the flow is low, check for open bypass or spill back valves. If the flow is too high, check the compressor speed.
- **Suction and discharge pressure.** Are the pressures normal? If not, begin to troubleshoot the system to find out what has changed. Look for a plugged strainer, fouled cooler, or incorrect valving.
- **Suction and discharge temperatures.** Are the temperatures normal? If the suction temperature is normal, but the discharge temperature is high, you may be looking at an indication of internal wear. A higher than normal discharge pressure may be caused by a higher than normal differential pressure.
- **High discharge temperatures or surging conditions.** These may indicate fouling. Check temperatures on any product coolers in the compressed gas stream. Fouling can be checked by looking at the color of the back flush water coming out of the exchanger or by a lack of differential temperature on water and gas flows into and out of the cooler.
- **Oil pressures and levels.** Be sure to check all lubrications levels and pressures, including seal oil pressures and flows if applicable.
- **Gas Seal Panel.** In the compressor is using dry-gas seals, check seal instrumentation panels to ensure flows and pressures are normal.
- **Bearing Temperatures.** If bearing temperatures are higher than normal first check over the lubrication system to make sure flows and supply oil temperatures are normal. Determine if it is the thrust bearing or the journal temperature that is running unusually high.
- **Vibration**. Are vibration levels normal? If not, have someone from the vibration department analyze the situation to determine what may be wrong and how long the compressor can continue to be run. Be sure to

monitor the thrust position monitors. Remember that a failure in the thrust bearing will severely damage the machine beyond replacing the damaged bearing.
- **Power loading.** High power levels may be due to a higher than normal speed or heavier than normal gas. For example, a compressor that pumps hydrogen normally might experience driver overloading if it was started with nitrogen still in the system.
- **Signs of surging.** If surging is detected, and it is not too severe, try and find the source. It may be eliminated by increasing flow by opening a spillback line or by removing any possible restrictions. If the surging is very severe and there are no other in plant instructions, unload the compressor or shutdown immediately and investigate the cause of the surging. Surging will wreck the compressor severely should it continue.

Common problems to watch for:

- Low or high rotational speed (Unless motor driven)
- Low or high suction pressure
- Low or high discharge pressure
- Gas that is too light or too heavy (Especially during start up or upset conditions.)
- Internal wear due to loss of inter-stage seals
- Internal fouling
- Plugged suction strainer
- Discharge system restriction due to a partially closed valve, fouled piping or coolers. Watch for surging under these conditions.
- Power loading. High power levels may be due to a higher than normal speed, heavier than normal gas, higher than normal differential pressure or high suction pressure.
- Higher than normal balance line temperature.

Reciprocating Compressors

Reciprocating compressors (see Figure 5.9) represent a class of gas compressors that raises the pressure of a gas by means of a tightly fitting reciprocating piston inside a cylindrical chamber, reducing volume from intake to discharge, in order to generate flow. Check valves within the cylinder are used to control the flow of gas within the cylinder.

Figure 5.9—Reciprocating Compressor Cylinder Cross Section

Reciprocating compressor components include:

1. A **piston** with rings and rider bands, cross head, and piston rod, which moves back and forth with each rotation of the crankshaft. The rings act like seals to keep gas from bypassing the piston. Rider bands are wide ring shaped devices that support the piston in the cylinder and prevent the two from touching. Some reciprocating compressors are "single acting" and some are "double acting". Single acting reciprocating compressors compress gas only when the piston is moving in one direction. When the piston moves toward the cylinder head the gas is being compressed but when the piston is moving away from the head, suction is pulling gas into the cylinder.
2. Double acting compressor cylinders compress gas on both movements of the piston. When the piston is moving toward the head of the cylinder it is compressing gas as well as putting it into the process. At the same time the piston is on the suction stroke

of the crankcase half of the cylinder. When the piston begins to move toward the crankshaft the crankshaft half of the cylinder is on compression and the head end is on suction. The effect is that there is almost twice the output from a double acting cylinder as from a single acting one.

A compressor **cylinder** contains the internal pressure and supports the reciprocating piston. The cylinder has a head on each end with water passages to remove some of the heat of compression. The outer head may include something called a clearance pocket for adjusting compressor output. The crank-end head has high pressure packing that fits around the piston rod. Its purpose is to keep the pumped gas from leaking past the rod and into the distance piece.

3. A set of **suction and discharge valves**. Reciprocating compressor valves are essentially check valves. The suction valves permit flow into the cylinder but not back out, and discharge valves permit flow to exit the cylinder but not return back in. Suction and discharge valve failures are the most common component failure in a reciprocating compressor. Unloaders may be part of the suction valve assembly.
    i. There are three ways of controlling the output of a reciprocating compressor that is motor driven (constant speed): reducing suction pressure; adding a clearance pocket to the cylinder by opening and closing the valve that separates the cylinder from the additional volume in the clearance pocket; activating unloaders on the intake valves, which will hold them open.
4. A **crank shaft, crank case, and connecting rods**. Some reciprocating compressors have two lube systems, one for the crank shaft and cross head and another for the cylinder and packing. The lubricant may be the same in both lubricating systems, but in some cases is different. Be careful not to mix or switch the two oils if they are different.

If all the compression components are working properly, the flow through a compressor cylinder is controlled by 1) the compressor speed, 2) the displacement of the piston within the cylinder, and 3) the suction pressure. Increasing either the compressor speed, cylinder displacement, or suction pressure will increase the net flow through the cylinder and the horsepower required from the driver. However, the design flow through a compressor cylinder can be reduced by numerous factors, such as internal leakage at the piston rings or valves, preheating of the inlet flow because of high ambient temperatures, a fouled exchanger on the suction gas, a lower suction pressure, etc.

Reciprocating compressors are often equipped with valve unloaders and clearance pockets as a means of controlling compressor throughput. If these flow control components malfunction, compressor flow may decrease dramatically or even drop to zero. Unloaders are control elements designed to hold intake cylinder valves open when they are activated, so that no gas can be trapped or compressed inside the cylinder during the compression cycle. Therefore, any compressor cylinders with activated unloaders will not able to contribute any flow to the compressor's total flow. The more unloaders that are activated the lower the total flow will be. Unloaders may be operated independently or in groups to produce the desired number of compressor "load steps." In contrast, clearance pockets are used to add or remove internal volume within a compressor cylinder, which in turn adjusts when the intake valves open and varies throughput. Generally, as flow is decreased with the use of unloaders and/or clearance pockets, the required horsepower decreases. Conversely, as flow is increased with the use of unloaders and/or clearance pockets, the required horsepower increases.

Troubleshooting Tips:

- If the differential pressure (i.e. the difference between the discharge pressure and the suction pressure) across any

compression stage increases, the discharge temperature will increase on that stage. Conversely, if the differential pressure across any compression stage decreases, the discharge temperature will decrease on that stage.
- If the compressor speed increases, the net flow will increase. The horsepower required by a compressor increases at higher speeds, which should be reflected by increasing amps or kilowatts. Conversely, if the compressor speed decreases, the net flow will likely decrease. The horsepower required by a compressor at lower speed will decrease, which should be reflected by decreasing amps or kilowatts. This relationship between compressor speed and horsepower is not seen on motor driven compressors unless the motor is equipped with a variable speed electrical drive.
- If the molecular weight of the gas being compressed increases, the required horsepower will also increase. Conversely, if the molecular weight of the gas being compressed decreases, the required horsepower will decrease.
- If the suction pressure increases, the mass flow and required horsepower will also increase due to an increase in the gas density. Conversely, if the suction pressure decreases, the mass flow and required horsepower will decrease. If a higher suction pressure puts more gas into the compressor, more horsepower will be required to achieve the same pressure.
- If a multistage compressor begins to lift a relief valve between stages, the stage downstream of the relief device needs to be checked as it is not "keeping up" with the previous stage.
- If throughput falls quickly, check to insure all unloaders on intake valves are functioning correctly and not allowing the intake valves to close. If clearance pockets are used to control compressor throughput normally, insure the valve that controls their addition or removal from the cylinder volume are closed to separate them from the cylinder.

These relationships will help to troubleshoot reciprocating compressors that are not performing as expected. When reporting a problem with a compressor, always list the symptoms that caught your attention, such as a low flow, high amps, knocking sound, vibration, not enough pressure, etc.

Reciprocating Compressor Start-ups:
Most large reciprocating compressors have a jacking or barring gear to allow the compressor to be turned over to insure that there is no liquid in any of the cylinders prior to starting. The compressor should be barred over at least twice before attempting to start a compressor. Reciprocating compressors are normally started with an open spillback line to minimize the process load on the driver. This method allows the circulation of some gas, while rotating the compressor and driver under a light load. Keep in mind that every compressor is unique in its design and construction, and therefore should have its own customized procedure to ensure all necessary steps are followed during start-ups. Only approved procedures that follow the manufacturer's recommendations and experience with this compressor in this application should be used—attention to detail is paramount.

Once full load conditions (i.e. speed, temperature, pressure, and flow) are reached, use the "Reciprocating Compressor Checklist" below to monitor machine condition and performance. Continue to monitor compressor conditions for at least an hour to make sure that everything is normal. If problems are encountered during this start-up phase, either call in help for a second opinion or shutdown until additional technical assistance is available.

Reciprocating Compressor Checklist

- Net flow through the compressor. If the flow is low, check for open bypass valves and, if unloaders are used, if they functioning correctly. If the flow is

too high, check the compressor speed and suction pressure.
- Suction and discharge pressure. Are the pressures normal? Remember, when troubleshooting, check the easy things such as valve positions, compressor speed, etc. first. Ideally there are pressure gauges on the suction and discharge pressure gauges and in-between stages.
- Suction and discharge temperatures. Are the temperatures normal? If not, begin to troubleshoot the system to find out what has changed. If the suction temperature is normal but the discharge temperature is high, you may be looking at an indication of internal wear or fouling valves. Use an infrared gun, contact pyrometer, or camera to check for hot valves.
- Oil pressures and levels. Check both crankcase oil and cylinder lube in case they are different. Watch for increasing crankcase levels which normally means water is entering the lubricant.
- Bearing temperatures. Are the bearing temperatures higher than normal? If temperatures are found to be elevated, first check over the lubrication system to ensure flows are normal and that supply oil temperatures are normal.
- Vibration of piping, compressor casing, or skid vibration. Are vibration levels normal? If not, have someone from the vibration department analyze the situation to determine what may be wrong.
- Packing leakage. Check for high pressure and ensure that distance pieces are not filling with liquid. Make sure liquid traps are working properly.
- Drainage traps. If there are "knock outs" between stages with traps for draining them, notice how much liquid is draining. If it is possible that liquid is entering a reciprocating compressor cylinder it can wreck the compressor completely in a very short time.

Common causes of problems to watch out for:

- **Abnormal rotational speed** (This is generally not a problem if the compressor is motor driven because motors are constant speed devices.)
- **Abnormal suction pressure**. Low pressure can be caused by a partially opened suction valve or by blockages up-stream of the first stage. High suction pressure can be caused by a control valve problem that regulates the pressure to the compressors first stage.
- **Abnormal discharge pressure**. A high discharge pressure can be caused by a blockage down-stream. A low discharge pressure can be caused by one of the preceding stages not compressing properly or low inlet pressure to the first stage.
- **Abnormal gas density**—especially if process upsets that can change the gas composition have been known to occur during previous start ups.
- **Internal gas slippage** due to leaking piston rings
- **Plugged suction strainer** that can result in a drop in flow or a loss of pressure.
- **Discharge restriction** due to downstream plugging or a partially closed valve.
- **Broken valves** will be evident by being much hotter than other valves. This condition can be caused by a bad valve with broken internals, a broken valve gasket, or a bad seat.
- **Fouled compressor valves** due to dirty gas will generally affect the output of the cylinder with the fouled valves.
- **Unusual gas temperatures** due to a loss of cooling somewhere in the process, usually caused by fouling or obstruction. Check intercoolers and after coolers on both the process side and water side.

Criticality

Compressors of all classes and designs should be monitored closely because they tend to be high-energy, high-horsepower machines. In many cases, they are unspared, which makes them highly critical to the processes they support. When they fail management takes notice.

Because compressors tend to represent the most critical machines at most sites, it makes sense to watch them closely and keep them maintained and operating efficiently. Here are just a few ways to ensure your compressors perform safely and reliability day in and day out:

1. Perform daily inspections
2. Check compressor performance on a regular basis. Keep process logs in order to know what is normal. Look at the logs; historical process logs will help you know how your compressors were performing yesterday, last week, and last month.
3. Analyze the lubricating oil regularly by checking levels and color of oil. Increasing oil level, in the crankcase, can mean water building in the crankcase.
4. Watch rod drop indicators and other indicators of potential problems on a daily basis.
5. If a divider block is used for lubrication, check for any indicator pins that may be popping out, indicating blocked lubrication to a cylinder or packing.
6. If there is a panel for high pressure packing leakage, monitor it for increased flow. Check flow on sweeps of distance pieces.
7. Operators should help perform regular maintenance as prescribed by the manufacturer when appropriate or during turnarounds in order to learn more about a compressor's internal construction. This type of hands-on knowledge is invaluable when troubleshooting and can make you feel more comfortable when inspecting and operating these machines.

Chapter 5
Review Questions:

1. A _____ is a fluid handling machine that takes gas at a lower density and pressure (in the suction nozzle) and compresses it by performing work on the gas so that it reaches a higher density and pressure (in the discharge nozzle).

2. Name two key reliability indicators for compressors.

3. What are the two main compressor design categories?

4. _____ compressors work by accelerating gas with a rotating impeller or set of rotating blades and then converting the exiting high velocity gas stream into a higher pressure by decelerating the gas stream for the next compression stage.

5. _____ compressors work by continually forcing gas into a smaller and smaller volume, either with a piston or with tight fitting rotors, and then expelling the reduced compressed volume into a discharge passageway.

6. List at least four (4) common causes of centrifugal compressor problems.

7. List at least four (4) common causes of reciprocating compressor problems.

Chapter 5
Answers:

1. A <u>compressor</u> is a fluid handling machine that takes gas at a lower density and pressure (in the suction nozzle) and compresses it by performing work on the gas so that it reaches a higher density and pressure (in the discharge nozzle).

2. <u>Temperature</u> and <u>vibration</u> are two key indicators of compressor health.

3. What are the two main compressor design categories?
   a. Dynamic compressors
   b. Positive displacement compressors

4. <u>Dynamic</u> compressors work by accelerating gas with a rotating impeller or set of rotating blades and then converting the exiting high velocity gas stream into a higher pressure by decelerating the gas stream for the next compression stage.

5. <u>Positive displacement</u> compressors work by continually forcing gas into a smaller and smaller volume, either with a piston or with tight fitting rotors, and then expelling the reduced compressed volume into a discharge passageway.

6. List at least four (4) common causes of centrifugal compressor problems.
   a. Low or high rotational speed
   b. Low or high suction pressure
   c. Low or high discharge pressure
   d. Gas is too light or too heavy
   e. Internal wear, due to loss of inter-stage seals
   f. Internal fouling
   g. Plugged suction strainer

h. Discharge system restriction due to a partially closed valve.
   i. Power loading. High power levels may be due to a higher than normal speed, heavier than normal gas or higher than normal differential pressure.

7. List at least four (4) common causes of reciprocating compressor problems.
   a. Low or high rotational speed
   b. Low or high suction pressure. A low suction pressure can be caused by a partially opened suction valve.
   c. Low or high discharge pressure. A high discharge pressure can be caused by a partially open discharge valve.
   d. Gas is too light or too heavy
   e. Internal wear due to leaking piston rings
   f. Plugged suction strainer
   g. Discharge system restriction due to downstream plugging or a partially closed valve.
   h. Broken or fouled compressor valves

Chapter 6
Lubrication Advice for Operators

By Drew Troyer

8 ways operators can ensure the effectiveness of a lubrication program:

- **Check levels in all reservoirs**—This is most important for small reservoirs such as overhung pumps and motor bearings that do not have an oil circulation system. Watch for increasing oil levels in equipment. In *turbines* it can mean steam seals are leaking, allowing the steam to enter the bearing housing and condense. In *reciprocating compressors* rising levels mean water is entering the crank case and accumulating. Since water is in the bottom of the reservoir and the lubricating pumps suck on the bottom of the reservoir or crank case, water will be circulated like oil and cause shorter machine life.
- **Check oil color**—If the lubricant is no longer the color it was when first installed the question "why" should be asked. If the answer is not readily available then

contact someone with more expertise to solve the problem.
- **Paper test**—put a drop of oil onto a paper towel or napkin; the oil will be wicked away. If there silver or golden specks are visible, something metallic is coming apart. The presence of metallic particles could be due to a failing bearing or severe rubbing taking place inside the machine. If water droplets are standing on the towel there is water in the reservoir. Look for a change in color from when the oil was new.
- **Smell the oil**—See explanation below in the section "The nose knows".
- **Check seals.** Oil reservoirs must be closed to outside elements, preventing dust or contaminants from entering. If there is a cap, make sure it is in place and tight with a gasket. If there is a flanged opening, check that it has a gasket and is sealed tightly

Figure 6.1 Desiccant appears to have reached half its useful life. Notice, the missing attachment bolts and gasket.

- **Check desiccant breathers** on reservoirs. They must have active material inside so they can continue to remove moisture from the reservoir (see Figure 6.1). Most manufacturers have a color code, similar to the yellow, green, black strip seen in Figure 6.1, that will determine when the desiccant material needs to be replaced.

- **Check containers with lubricant or a barrier fluid** reservoir making sure they are well sealed and that automatic closing mechanisms are working properly. This ensures that rain or steam from the area cannot enter the fresh oil (see Figure 6.2). If you must keep a container near a piece of equipment to fill the reservoir often, tell someone of the problem so they determine the cause of the excessive oil usage. This can create a housekeeping problem as well.

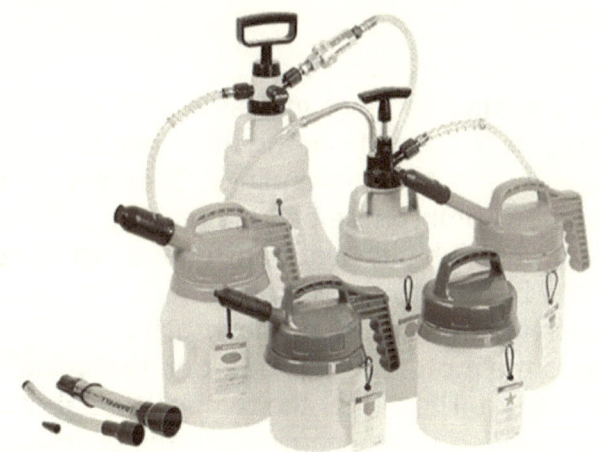

Figure 6.2—Oil containers designed for sealing against the elements and controlled pouring of lubricants.

- **Know the correct lubricant** to add to a reservoir when it needs to be topped off. Marking the type on the reservoir can prevent adding the wrong kind.

- **Improving Electric Motor Lubrication Policies**
  Electric motors—more specifically, grease-lubricated electric motors—are the prime movers for most machines in plants and factories. Motors that are larger than 25 horsepower require occasional

regreasing. Doing so properly is essential to maintaining equipment reliability. Upon inspecting poorly lubricated motors, we often see signs of over greasing, under greasing, caked-up thickener suggesting dry-out or too long between grease additions, and a "rainbow" of colors, indicating that the motor is being greased with whatever product is handy. All of this indicates a lack of control over the motor lubrication process.

Here are some good reasons to improving your motor lubrication techniques if you are responsible for their lubrication.

- **Facts and Observations**
  - **Over greasing** bearings increases heat and leads to separation of the oil from the thickener, as well as to grease getting into the windings. Overheating and separation compromise bearing reliability. Grease in the windings inhibits heat transfer and increases the rate at which motor insulation material degrades. An increase in temperature of the insulation on a motor reduces the motor life in proportion to the high temperature. The higher the temperature above that recommended on the name plate, the shorter the motor life will be, which is why motors should not be run overloaded or within their stated service factor. (A motor's stated service factor (*SF)* is an indication of temporary overload capability that will not result in damage.)
  - **Under greasing** the motor reduces film thickness, which increases frictional heat in the bearing. This causes the oil to separate from the thickener, which leads to caking. Under greasing can cause the bearing to run in the boundary layer regime and the bearing life will be shortened.
  - **Mixing greases** can lead to a range of problems. First, counter to conventional wisdom, it is not true that

"grease is grease." Grease is a complex combination of chemicals—base oil, additives, and thickener. Motor bearings require the right combination of performance characteristics for the application. Failure to deliver these properties compromises motor reliability. Also, grease components don't always play nice with one another. Incompatibility can cause separation of the base oil and thickener, sludge, varnish, and compromised lubrication performance.

- **Using the wrong grease** is frequently selected. Multipurpose greases can reduce motor life. First, there is no particular standard for "multi-purpose grease". Its base oil is typically higher than what's required for an electric motor application, which wastes energy and unnecessarily generates heat. It may or may not utilize a thickener that's appropriate for the application, and multi-purpose grease frequently contains an extreme-pressure (EP) additive that can cause corrosion of the bearing cage and on windings in the event of over lubrication.

The Nose Knows—Using Odor as a Test for Your Oils

Odor, like color, clarity, presence of foam and other characteristics of the oil that can be evaluated with the human senses, can effectively identify degradation or contamination of lubricating oils. As simple as it seems, smelling oils should be a regular part of routine machine inspection process. A lot of caution should be exercised when smelling oil samples. Do not place the sample directly under your nose—remember that there may be toxic chemicals in the sample from the process. Rather, wave your hand above the opening to waft the scent toward your face. Sometimes it helps to heat the oil, or take a "live" sample as from a dip

stick, which increases the likelihood of detecting certain contaminants and degradation by-products.

Odor assessment, when used often, will train the nose to detect many potential problems with oil. Contaminants such as ammonia, chlorine and other chemicals are easily detected and generally affect the lubricant in a negative way.

Typical odors associated with lubricant problems you can expect to find:

- **Oxidation**—sour or pungent odor, acrid (rotten egg) smell, or something similar to stale cheese
- **Thermal Failure**—smell of burnt food or burned oil
- **Bacteria**—stench, road-kill smell
- **Contaminants**—solvents, refrigerants, degreasers, hydrogen sulfide, gasoline, diesel, kerosene, and process chemicals will have the lubricant smell of the specific contaminant.
- **Amino Acids**—fish odor
- **Nitro Compounds**—almond-like scent
- **Esters (Synthetic Lubricants) and Ketones**—perfume odor
- **Chemicals**—odor of the chemical being pumped or compressed

Chapter 6 Questions

1. List 5 of the 8 ways operators can ensure the effectiveness of a lubrication program.

2. _____ electric motor bearings increases heat and leads to separation of the oil from the thickener, as well as to grease passing the labyrinth and getting into the windings.

3. How can you tell when the desiccant on the reservoir breather needs to be changed?

4. List three smells of lubricant that can tell something about its condition.

5. List two ways to know what the correct oil to add to a reservoir is.

6. Describe what a good container for oil should have.

Chapter 6
Answers

1. List 5 of the 8 ways operators can ensure the effectiveness of a lubrication program:
   - Check levels in all reservoirs
   - Check oil color
   - Use the paper towel test
   - Smell the oil
   - Always insure oil reservoirs are closed to the outside elements preventing dust or contaminants entering.
   - If there are desiccant breathers on your reservoirs, ensure they still have active material inside, so they can continue to remove moisture from the reservoir.
   - If there are containers with lubricant to add to a piece of equipment or barrier fluid reservoir ensure the container is well sealed or of the automatic closing type to prevent rain or steam from entering fresh oil.
   - Insure you know the correct lubricant to add to a reservoir if you are supposed to top it off.

2. **Over-greasing** electric motor bearings increases heat and leads to separation of the oil from the thickener, as well as to grease passing the labyrinth and getting into the windings.

3. How can you tell when the desiccant on the reservoir breather needs to be changed?
   A change in the color of the desiccant is an indication that is needs to be replaced. There is usually a color code indicator on the breather designating when the desiccant is depleted.

4. List three smells of lubricant that can tell something about its condition.
    - **Oxidation**—sour or pungent odor, acrid (rotten egg) smell, or something similar to stale cheese
    - **Thermal Failure**—smell of burnt food or burned oil
    - **Bacteria**—stench, road-kill smell
    - **Contaminants**—solvents, refrigerants, degreasers, hydrogen sulfide, gasoline, diesel, kerosene, and process chemicals will have the lubricant smell and the specific contaminant.
    - **Amino Acids**—fish odor
    - **Nitro Compounds**—almond-like scent
    - **Esters (Synthetic Lubricants) and Ketones**—perfume odor
    - **Chemicals**—odor of the chemical being pumped or compressed

5. List two ways to determine the type of oil that should be added to a reservoir.
    - Look for a marking on the reservoir; look for any punched or etched metal tags that are attached to the reservoir, or look in your CMMS (computer maintenance management system) software system under the machine details for lubrication recommendations.

6. Describe what a good container for oil should have. A good container should contain the following features: It should be completely enclosed, waterproof, and color coded for the required lubricant. The oil container should also be designed to allow you to easily view the condition of the oil and its oil level

Chapter 7
More Machinery Best Practices
by Robert L. Matthews, Reliability Manager for Royal Purple, LLC

The Devil is in the Details
"The devil is in the details" is an adage applicable to machinery maintenance. It means that reliability requires close attention to key machine details, such as design, assembly, installation, and maintenance. But, if the operation of equipment is not right, these other items cannot improve reliability. For example, a pipe hanger located 20 or 30 feet from a pump that is not functioning properly can have a significant effect on pump nozzle loads. Consequently, if the pipe hanger fails or malfunctions further, pump failure can soon follow. While a pipe hanger may not seem to be machinery or operations related, a failed pipe hanger should be considered a critical component failure in the

pump-foundation-piping-control system. If detected early, the problem can be corrected.

If you want to set yourself apart from the others in your group, learn all you can about machinery and machinery systems. A simple change in perspective can have a major impact on how you view and understand your machinery. By widening your view of machines and thinking about them in terms of integrated systems, you can begin to understand how critical internal and external components must interact with one another to provide a reliable system. Operators play an important role in ensuring the ongoing reliability of rotating equipment by closely monitoring them with a trained eye. They can detect problem that if addressed quickly can mitigate or eliminate future problems.

In this chapter, you will find valuable insights about key machine components, such as anchor bolts, pipe hangers, couplings and how they can affect the overall reliability of machines. Also covered is the importance of knowing the machinery design and operating details when troubleshooting process machines.

Anchor Bolt Inspections

Figure 7.1—Centrifugal Pump and Motor Attached to Concrete Foundations with Anchor Bolts

Anchor bolts play a vital role in the restraint, support, and alignment of process machinery. Broken or loose anchor bolts will lead to excessive vibration or allow machine-to-driver misalignment.

Here are some proven tips for maintaining the effectiveness of anchor bolts:

- Anchor bolts must be tight. Look for indications of looseness such as liquid being pushed in and out from under the bolt, washer, or frame of the equipment. Occasionally try to turn anchor bolt nuts by hand or with a tool to see if they are tight.
- Tap on the anchor bolt to listen for a dead sound that might indicate a broken bolt. See if the nut and anchor in the foundation move. If pliers are used to test tightness, a slight pull upwards on the bolt or stud will help to indicate if it is anchored or has come loose.

Looking Up
How often do you look upward while walking through your plant? Do you ever inspect the pipe, pipe supports, spring cans, bolting, insulation, and so on? This inspection technique is a must for safety and equipment reliability. I remember seeing a nut on the floor of a plant and looked up to see that it came from a 4 bolt flange. The flange had only 3 bolts and was labeled "Sulfuric Acid." Consider how many times a day someone walked under the potentially leaky flange.

Unmaintained supports are one of the largest, if not the largest, contributors of maintenance issues we see today. (Figures 7.2 through 7.4 are examples of unmaintained piping supports.) Failed pipe supports lead to high nozzle loads, which put added loads on bearings and seals and eventually cause premature equipment failures. Furthermore, when equipment bases are loosened due to piping strain, often the base is repaired without a resolution of the true cause of the problem. Unresolved piping strain will continue to shorten the life of the grout, causing base looseness and related misalignment issues.

Pipe supports in hot service and around steam turbines are particularly important. Failure in any way of these supports or slides that allow controlled movement can result in premature equipment failure.

Figure 7.2—Notice that one valve support is not contacting pipe

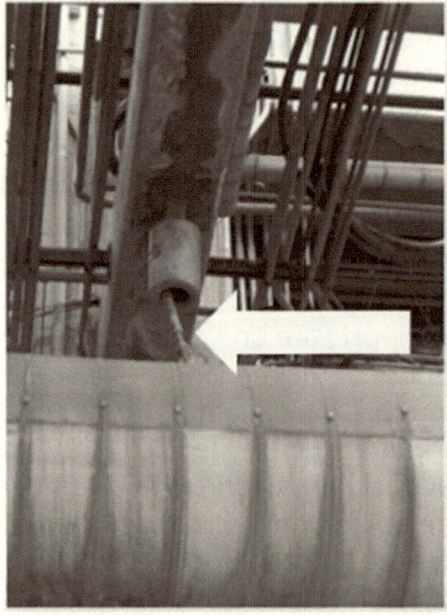

Figure 7.3—Hanger support is broken

Figure 7.4—A rope support has slipped and cocked the pipe

Coupling Inspections

When passing by machines that are coupled, look for parts on the ground that may have come from a separating coupling. Items may include broken shims from a shim-pack coupling. You should also look for the dust that is created by the degradation or severe misalignment on couplings with a nonmetallic coupling spacer. A strobe light can be used to inspect the coupling while the machine is running, provided the coupling guard allows it. Operators should always ensure that maintenance crews clean up after their work, so that any new material found on the pedestal, equipment base or around the machine is from the installed coupling.

Know your Equipment

Know your equipment. Have you heard complaints that a pump is not producing enough flow or pressure or that a gear box is running too hot? The first question we need to answer is: What was the flow, pressure, or temperature yesterday or last week when there was no problem? Only operators can answer those types of questions. Key operating conditions should be recorded after starting any piece of critical equipment and should be taken and recorded on a regular basis.

It is also important to know the basic construction of your equipment. Figure 7.5 is an example of a gear box schematic that may help troubleshoot when a problem is detected.

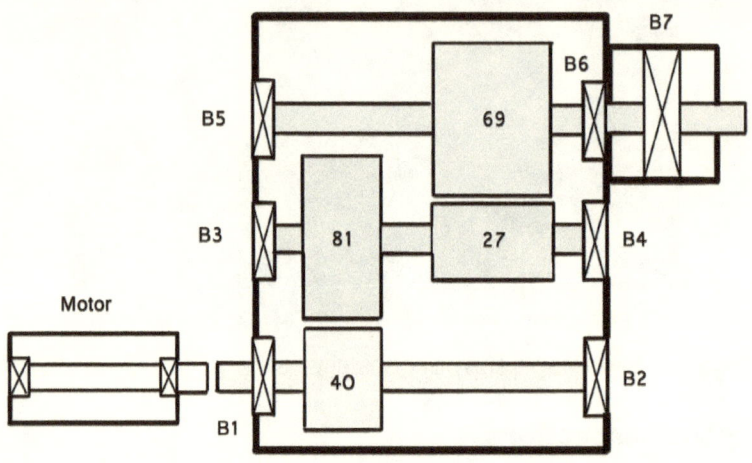

Figure 7.5—Gearbox Field Sketch

Always note current ambient temperature when evaluating equipment temperature readings. If the ambient temperature is substantially higher or lower than when the readings were first recorded, it may be necessary to take that factor into account when looking at the current value. Looking for obvious changes in any of the operating conditions is a simple, yet effective method of field troubleshooting.

Remember that capturing actual field data with an accurate instrument is invaluable when troubleshooting. If someone thinks the gear box is hot because it's running 160°F when it was running 150°F at startup, you have data to prove whether it is or isn't. It is more helpful to have objective values, such as a bearing housing reading of 150° F, than to have subjective information such as "hot to the touch." Your senses are a good indicator to know when it is necessary to use an objective measuring tool.

Walk-Through Secrets: *What to Look For On Your Regular Inspections Rounds*

Walking through a plant and looking at the equipment can be an eye-opening experience, because there are many opportunities in the plant to prevent catastrophic damage to equipment. Failure detection and prevention starts with knowing what "normal" is for your equipment. Many small issues can be fixed or corrected that will eliminate large safety, environmental, and costs in the near future. Remember, there are many abnormal conditions or problems that operators see that no one else does. In addition, the operator is the only one that knows what normal sounds like, looks like, or feels like. For example it is usually not good to see foam in the lubricating oil. If, as the operator, you saw foaming yesterday, last week, last year and for the past five years and the equipment has had no repairs during this time then the presence of foaming in this machine, in this process, in your plant is likely not a problem. Only an alert operator would know this type of equipment-specific information and could use this valuable resource should a problem arise on this piece of equipment.

Figure 7.6 shows two oil bowls, one for the turbine and one for the pump. What should catch your attention is that the two oil colors are very different. If, as an operator, you do not understand why this is happening, this condition should be brought to someone's attention. Is the dark color on the turbine because the oil has become too hot and oxidized? Is the oil on the pump freshly changed? Are the two oils simply different? Perhaps, the oil in the pump appears light when new and the turbine oil appears darker when it is fresh. These are just a few lubrication observations to consider when making the rounds through the plant.

Figure 7.6—Oil bowls on a centrifugal pump with turbine driver

The equipment in Figure 7.7 has several problems, all of which should catch your attention. The first problem is that the seal is missing a bolt, which should be a big concern. Another issue is that the pressure gauge is rusty, has no gauge glass, and is pegged on the zero stop.

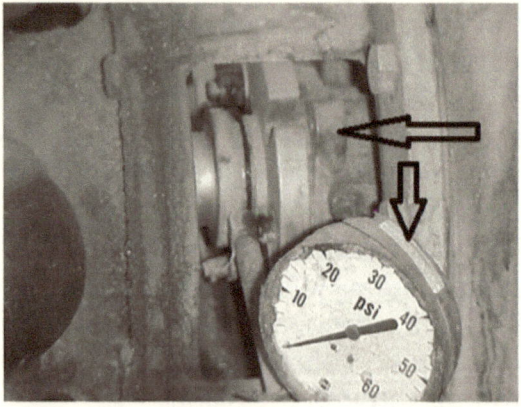

Figure 7.7—Pump with several problems

The mechanical seal in Figure 7.8 is installed on a boiler feed pump. This should be an API (American Petroleum Institute) seal flush plan 21 that is composed of a heat exchanger on the stuffing box flow to cool the water entering the seal area. If the exchanger is not working, the additional heat generated by the seal can cause flashing or boiling on the seal faces which will be the same as the seal running dry. The operator should check the exchanger by detecting a differential temperature to and from the seal as well as inlet and outlet water of the exchanger. This can be done with tools or by touch. Precautions should be taken to ensure you are not burned.

Figure 7.8—Seal in a boiler feed pump

Mechanical seals require start-up attention and often go unchecked. Inspection at start-up is imperative to ensuring the necessary flush rates are applied if applicable.

The expansion joint shown in Figure 7.9 is obviously over-compressed. This is a catastrophic release waiting to happen. Some expansion joints have "stay rods" (see the expansion joint with stay rods in Figure 7.10) that must be connected. Stay rods limit the amount of expansion that can take place and prevent the expansion joint from blowing up like a balloon and pulling apart.

Figure 7.9—Over-compressed expansion joint

Figure 7.10 Expansion joint loose stay rods

Take the time to inspect pumps, piping, supports, lubrication, and all associated equipment at your plant.

One new technology that can be used to look at your running equipment and flow process piping is the infrared camera, which uses a technique known as thermography. The camera shows heat differences inside piping and equipment that can indicate restrictions, hot spots, and other problems. Infrared cameras can be used on motors, pumps, turbines, fans, bearings, compressor valves, and more as a new way of predicting failures. The first indication of a machinery problem can often pay for the cost of the camera many times over if a failure is averted.

The thermograph in Figure 7.11 shows that the lip seal in an ANSI pump is running about 100 degrees hotter than any other part of the pump. This is why so many smart maintenance groups have followed the API and ANSI recommendations that labyrinth seals, not lip seals, be used on bearing housings. The higher the speed, the hotter the lip seal operates when it is newly installed; then as the seal wears, the heat is reduced, and the seal's ability to seal is lost.

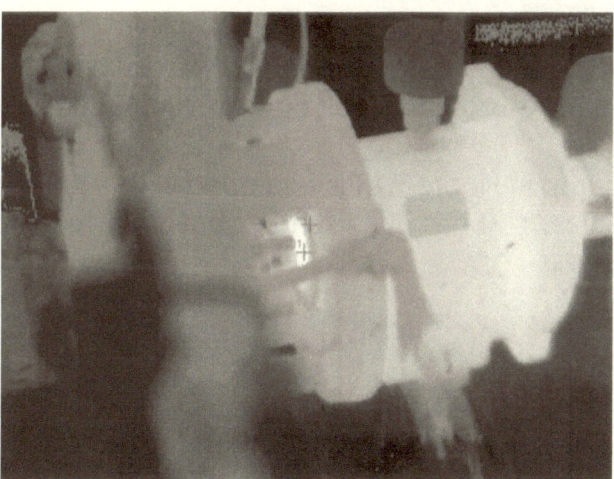

Figure 7.11—Thermograph showing a hot lip seal

Figure 7.12 shows a centrifugal pump casing foot with a loose bolt. The bolt is not doing its job of securing the pump case to the baseplate and maintaining its alignment. Whether the bolt is a little loose or very loose, the effect on alignment is the same. One condition is just more obvious than the other. When you find these types of problems, document the information so it can be used in a Root Cause Failure Analysis (RCFA) if required. In many instances, loose hold-down bolts go undetected or undocumented and cause mechanical seal or bearing failures. And when the equipment ultimately fails, the seal or bearing will likely be wrongly blamed allowing the potential for the problem to reoccur.

Figure 7.12—Loose foot bolt on a centrifugal pump casing

The pump in Figure 7.13 is equipped with a bull's-eye sight glass that has been sand or bead blasted and now it's useless.

Figure 7.13—Damaged bull's eye sight glass

If there is a clear plastic drain installed on your equipment as in figure 7.14 watch it closely. With this type of trap in place, you can easily see water in the bottom of the pump housing you can detect wear particles and oil color changes and bleed off any water contamination once detected.

Figure 7.14—Oil sight on sump

Figure 7.15 shows a pipe shoe that does not contact the pipe's support, which causes pipe strain, a condition where piping attached to machines induces unacceptably high stress into the machine frame and support system. The worst case would occur if the pipe shoe moved so far that it hooked over the support and could no longer move at all.

Figure 7.15—Pipe shoe not contacting pipe support

Summary:
This chapter can be summarize by stating that to maximize the reliability of your rotating equipment you should 1) know your machines, 2) follow your machine-specific procedures, 3) be vigilant of changes, and 4) perform regular walk-throughs or rounds. The observations and data recorded during your regular rounds will be invaluable if a root cause failure analysis is ever performed on one of your failed rotating equipment. No machine detail or observation is too insignificant when it comes to equipment safety or health.

Chapter 7 Questions

1. Occasionally try to turn _____ nuts by hand or with a tool to insure they are tight.

2. Failure detection and prevention starts with knowing what _____ is for your equipment.

3. List four ways to maximize the reliability of your rotating equipment.

4. List 2 items that are external to your machine that, when not installed properly or maintained, can reduce the life of your equipment.

5. List at least two things to look for during a coupling inspection.

6. If supervision does not act on your findings, should you stop telling them of problems you have found?

Chapter 7
Answers

1. Occasionally try to turn **anchor bolt** nuts by hand or with a tool to insure they are tight.

2. Failure detection and prevention starts with knowing what **"normal"** is for your equipment.

3. List four ways to maximize the reliability of your rotating equipment.
   - Know your machines.
   - Follow your machine-specific procedures.
   - Be vigilant of changes.
   - Perform regular walk-throughs or rounds.

4. List 2 items that are external to your machine that, when not installed properly or maintained, can reduce the life of your equipment.
   - Pipe supports
   - Flex legs
   - Pipe expansion joints
   - Expansion slides

5. List at least two things to look for during a coupling inspection.
   - Coupling dust on the baseplate from a flexible coupling
   - Broken coupling spacer shims
   - Broken coupling bolts
   - Coupling keys
   - New dents in the coupling guard.

6. If supervision does not act on your findings, should you stop telling them of problems you have found?
   No, you should not stop telling supervision about problems you have found. It is the only way your observations can ever be acted on

Closing Thoughts

The goal of this book was to present: 1) Factors that are essential to healthy equipment operation and longevity; 2) Proven machinery field inspection methods for process operators and operating personnel. To reach this goal, we briefly covered the construction of some of the commonly used process machine types, how they function, and how to utilize human senses to evaluate their health. The material covered in this book was meant to be an introductory course that can provide a basis for more advanced learning.

The simple techniques of touching, listening, and visually inspecting machinery while on rounds or passing by equipment will ensure the best possible life for machines and reduce the likelihood of unexpected failures. Readers probably already use the inspection methods presented here while in or around their homes. We all listen to our autos, washing machines, and refrigerators. We use our sense of smell to check if something is burning or overheating. These types of inspections are nothing new to us. The challenge is to incorporate them effectively into daily routines while inspecting critical machines.

The authors sincerely hope they succeeded in presenting useful information for process operators and that your company benefits greatly from the application of the material presented here. The overall effectiveness of this training will ultimately be measured by increased equipment reliability and lower repair costs.

Appendix A

Useful engineering facts
1. When something gets hotter, it will generally expand.
2. When something gets colder, it will generally contract.
3. If you bend metal, such as a coat hanger, back and forth, it will get warm and finally break, that is, it will become fatigued and fail.
4. Atmospheric pressure is approximately 14.7 psia (pounds per square inch absolute) at sea level.
5. Zero (0) absolute pressure (0 psia) is what is found in outer space, which is nearly a perfect vacuum. In other words, 0 psia is the absence of all pressure.
6. 0 gauge (0 psig) is 14.7 psi absolute. If you take a gauge that reads 0 on earth and sent it into outer space, it would read negative 14.7.
7. A column of water approximately 2.4 feet high has a pressure equal to 1 psig at the bottom of the column.
8. One psi of pressure is equal to approximately 0.43 feet of water.

Useful Conversions
When collecting and evaluating data on machinery condition, it is critical that units of measurement be carefully documented. For example, if the manufacturer recommends a maximum bearing temperature of 175 ºF, you must ensure that you record the reading in degrees Fahrenheit, converting centigrade readings if necessary, for a direct comparison. Similarly, if you are measuring vibration levels in inches per second, but the manufacturer provides vibration levels in millimeters per second, you must be able to convert the metric reading into English units in order to make a valid evaluation.

Always ask the following questions, to ensure you are using the proper measurements units in your evaluation:

- What are the units of the measurement I am taking?
- What is the measurement limit provided by the manufacturer or the engineering department?
- Do I need to change the measurement scale or do I need to make a conversion to compare my reading with the recommended limit?

If you are not sure about your measurement units, ask someone who can help.

Here are some conversion formulas and tables you might find useful when dealing with machinery:

Rotational Speed Conversions
Revolutions per minute = Revolutions per second x 60
Example #1: 60 revolutions per second x 60 = 3600 revolutions per minute
Rotations per second = Rotations per minute / 60
Example #2: 1800 revolutions per minute / 60 = 30 revolutions per second

Speed
Inches per second = Millimeters per second / 25.4
Millimeter per second = Inches per second x 25.4
Inches per second = Feet per second / 12
Feet per second = Inches per second x 12
Inches per second = Meters per second x 39.37
Meters per second = Inches per second /39.37

Distance and Length Conversions

Metric Conversion Factors	
1 centimeter =	10 millimeter
1 decimeter =	10 centimeter
1 meter =	10 decimeter
1 Angstrom =	10^{-10} meter

English & Metric Conversion Factors	
1 inch =	25.4 millimeter
1 inch=	25,400 microns
1 feet =	12 inches
1 yard =	3 foot
1 mil =	0.001 inch

Common Volume & Capacity Conversion Factors

1 cubic yard (cu yd.) = 27 cubic feet
1 teaspoon = 1/3 tablespoon
1 tablespoon = 1/2 fluid ounce = 3 teaspoons
1 U.S. fluid ounce (fl oz.) = 1/128 U.S. gallon = 1/16 U.S. pint
1 cup = 1/4 quart = 1/2 pint 8 fluid ounces
1 pint (pt.) = 1/8 gallon = 1/2 quart = 16 fluid ounces
1 quart (at) = 1/4 gallon = 32 fluid ounces
1 U.S. gallon (gal) = 231 cubic inches
1 milliliter (ml) = 1/1,0000 liter = 1 cubic centimeter
1 centiliter (cl) = 1/100 liter = 10 milliliters
1 deciliter (dl) = 1/10 liter
1 liter = 1 cubic decimeter
1 cubic inch = 16.4 cubic centimeters
1 cubic foot = 0.0283 cubic meter
1 cubic yard = 0.765 cubic meter
1 fluid ounce = 29.6 milliliters
1 U.S. pint = 0.473 liter
1 U.S. quart = 0.946 liter
1 U.S. gallon = 0.84 imperial gallon =3.8 liters
1 dry pint = 0.55 liters

1 dry quart = 1.1 liters
1 cubic centimeter = 0.06 cubic inch
1 milliliter = 0.034 fluid ounce
1 liter = 1.06 U.S. quarts = 0.9 dry quart

Temperature Conversions
Fahrenheit to Centigrade temperature conversion:
°F = (1.8 x °C) +32
Centigrade to Fahrenheit temperature conversion:
°C = (°F-32) x 0.555
Example #1: 212 °F = (212—32) x 0.555 = 100 °C
Example #2: 130 °C = (1.8 x 130) + 32 = 266 °F

Vibration Conversions
Inches per second = millimeters per second / 25.4
Inches per second = microns per second / 25,4000
Millimeter per second = inches per second x 25.4
Microns per second = inches per second x 25.400

Index

A

Advice   xv, 115
Anchor Bolts   71, 76, 78, 81, 88, 126-7
API plan 21   132
API plan 32   31
Audible   54, 58-9
Automatic grease lubrication   52
Axial   37, 94

B

Bearing types   18, 20
Bearings   xiv, 2-3, 17-20, 28-9, 31, 34, 36-9, 41, 45-6, 50, 52, 65, 84-5, 118-19, 151
Bernoulli's Principle   3, 6-7
Bernoulli's Principle Centrifugal pumps   3, 5, 8
Boundary lubrication   38-9

C

Causes of mechanical seal failures   27-8
Centrifugal compressors   xv, 2, 7, 36, 94-8, 151
Circulating   15, 43-5, 50, 52, 84
Compressors   xv, 2-3, 7, 31, 36, 70, 91, 93-9, 102-7, 109-11, 113, 115, 151
Couplings   1, 62, 126, 129, 151

D

Driven machines   1-2, 14, 34, 36
Drivers   xiv, 1-2, 12, 34, 36, 91

E

Electric motor inspections   86, 89
Electric motors   2, 12, 36, 62, 117

F

Forced   44-5, 52, 57
Functions   23, 37, 50, 52

G

Gear boxes   xv, 2, 12, 16, 31 70, 151
Gearbox inspections   76
General machine shut-downs   84
General machine start-ups   82
Grease   19, 39-41, 47-8, 52, 117-19, 121, 123

H

Hard hat   60
Heat exchangers   xv, 31-2, 151

I

Infrared camera   61
Infrared gun   108

Inspections   24, 26, 43, 48, 50, 53-5, 57-8, 67, 69, 72, 76-7, 110, 126, 129, 131

K

Key reliability indicators   96, 111

L

Lubrication   xiii-xv, 18, 21, 23-5, 37-48, 50, 52-3, 56-7, 76-7, 83-4, 110, 115, 117-19, 123-4, 153-4
Lubricators   47, 83-4

M

Mechanical seal flush plans   29
Mechanical seal leak paths   28
Mechanical seals   21, 23-4, 26-9, 36, 71, 133, 151, 154
Methods of lubrication   18
Multi-staging   95

O

Oil mist   45-7, 52

P

Packing   3, 21-6, 36, 71-2, 79, 104, 108, 110, 151, 154
Packing adjustments   25
Packing inspections   24
Packing Leakage Paths   25, 26
Packing versus mechanical seals   23
Pipe Hangers   126

Plain   18-20, 36
Positive displacement pumps   xv, 3, 10-11, 36, 70, 151
Program   xiii, 37, 115, 121, 123
Pump inspections   71, 86, 88
Pump seals   34
Pumps   xv, 2-3, 5, 7-8, 10-12, 15, 22-4, 26, 36, 44-5, 91, 96-7, 115, 134-5, 151-2

R

Reciprocating compressors   xv, 2, 95-6, 102-5, 107, 115, 151
Regimes   38
Ring lubrication   43
Rolling element   18-20, 36, 38-9, 41, 55, 69
Rotary   2, 12, 93-4

S

Seal flush plans   29-30
Smells   53, 121, 124
Speed modifiers   1-2
Splash lubrication   41, 43
Start-ups   82, 84, 100, 107
Steam turbine inspections   78, 86, 89
Steam turbines   2, 12, 14, 16, 36, 83, 91, 128, 151
Stethoscope   59
Strobe light   62, 129

T

Tactile 57-8
Temperature 14, 17, 27, 60-1, 63-5, 74, 77, 80-2, 90-2, 95-6, 99-102, 106-8, 118, 129-30, 146
Temperature measurement equipment 63-4
Tips on greasing 40
Trending 65
Troubleshooting iii, xiii, 8, 27, 64, 96, 98, 105, 108, 110, 126, 130
Types 1-3, 10, 12, 14, 18-20, 32, 34, 36, 39-40, 42, 46, 70, 93, 129, 142
Types of lubrication oils 42

U

Ultrasonic gun 59
Ultrasonics 59

V

Valve wrench 60
Vibration xii, 9, 12, 17, 26, 28, 57, 63-4, 66, 69, 72, 96, 101, 107-8, 143
Visual 47, 55, 58, 60, 62, 64-5, 67, 69, 73, 85

Meet the Authors

Mr. LeBleu has taken and taught many classes on simple vibration analysis, lubrication, alignment, reciprocating compressors, centrifugal compressors, fans, centrifugal pumps, positive displacement pumps and mechanical seals and packing as well as set up predictive and preventative maintenance programs. He is familiar with motors, generators and gas and steam turbines as well. His field of experience involves rotating equipment failures of all types such as pumps, motors, the mechanical portion of the failure, seals, bearings, turbines both steam and gas, fans, and couplings. He has over 35 years of experience on rotating equipment of most every kind.

Through his years of experience, Mr. LeBleu has organized predictive and preventative maintenance programs at major chemical plants, has conducted root cause analysis on equipment failures on steam turbines, compressors, pumps, motors, gear boxes, seals, couplings, bearings, heat exchangers, etc. and determined a solution to eliminate reoccurrences. He trained young engineers in alignment and various other disciplines necessary to maintenance and reliability engineering. He has written numerous procedures for the maintenance, operation, and overhaul of pumps, compressors and rotating equipment. He taught inspection of heat exchangers, columns, vessels, piping and valves to engineers and operators and also taught for two years in power generation at an Air Force technical school.

Julien LeBleu, Jr. and Robert Perez

Robert X. Perez has 30 years of rotating equipment experience in the petrochemical industry. He earned a BSME degree from Texas A&M University (College Station), a MSME degree from the University of Texas at Austin, and is a licensed professional engineer in the state of Texas. Mr. Perez served as an adjunct professor at Texas A&M University-Corpus Christi, where he developed and taught the *Engineering Technology Rotating Equipment* course. He has written numerous machinery reliability articles for technical conferences and magazines and served on the Turbomachinery Symposium Advisory Committee.

Mr. Perez has authored *"Operator's Guide to Centrifugal Pumps"* (Xlibris 2008), co-authored (with Andy Conkey) *"Is My Machine OK?"* (Industrial Press 2011), and is the technical editor of the *Illustrated Dictionary of Essential Process Machines Terms* (Diesel & Gas Turbine Publications Inc. 2013).

Meet the Contributors

Drew Troyer is a renowned maintenance and reliability educator and thought leader, with a passion for lubrication and the importance of lubrication to overall plant reliability. Widely published, Drew has authored more than 125 technical papers, articles, books and book chapters on industrial reliability management. Drew specializes in quantifying the financial benefits of plant reliability management and in conveying those opportunities to senior-level managers in the language and style to which they're accustomed.

He has developed dozens of proprietary processes and software applications to help define the strategic goals for managing plant reliability, economically justifying the plant's elements and successfully implementing the strategy. Drew is a Certified Reliability Engineer (CRE), a Certified Maintenance & Reliability Professional (CMRP), holds an MBA and a post-master's graduate study in reliability engineering technology management and measurement theory.

Julien LeBleu, Jr. and Robert Perez

Bob Matthews has 35 plus years reliability maintenance experience ranging from hands-on to supervision, plant maintenance management, consulting, and training. His training resume includes the development and presentation of courses on Advanced Pump Rebuilding, Lubrication, Mechanical Seals, Braided Packing, Operator Awareness, and other customized classes for diverse companies like Exxon, Phillips and Westinghouse as well as programs for Auburn, the University of Alaska, LSU, the Vibration Institute, ASME, FSA, and others.

Today, Bob is the reliability manager for Royal Purple, LLC where his role includes providing added value services to the maintenance industry and their customers with lectures, training, and plant visits for reliability recommendations. Bob is a graduate from Lamar University, and has authored several industrial publications.

www.ingramcontent.com/pod-product-compliance
Lightning Source LLC
Chambersburg PA
CBHW031053180526
45163CB00002BA/823